The Challenges of Democracy in the War on Terror

This book unravels the role of democracy after the 9/11 terrorist attacks and reflects important debates surrounding the security of Muslim communities in the years to come. It looks at the problems of torture, violence and the legal resources available to contemporary democracies to confront terrorism.

While terrorism is often regarded as one of the major threats to the West and the nation-state, this book explores the notion that a disciplined sense of terror is what keeps society working. The strengths and limitations of liberalism are examined, as well as the ethical dilemma of torture and human right violations in the struggle against terrorism. This book carefully dissects the origin of the nation-state and how it keeps society united.

The author offers a creative and unique approach to democracy and worldwide terrorism, exploring the consequences for the nation-state. This book looks at the connections between terrorism, mobility, consumption, torture and fear. It will be of interest to researchers as well as postgraduate and postdoctoral students within the fields of Human Geography, Politics, Media and International Relations.

Maximiliano E. Korstanje is Senior Researcher in the Department of Economics at University of Palermo, Argentina. He was awarded Visiting Research Fellow at the School of Sociology and Social Policy at the University of Leeds, UK, and at the University of Havana, Cuba. He was recently awarded Emeritus Chief Editor of the *Journal of Cyber Warfare and Terrorism*. His recent books include *The Rise of Thana-Capitalism and Tourism* (2016), *Terrorism, Tourism and the End of Hospitality in the West* (2017), and *Mobilities Paradox: A Critical Analysis* (2017).

Routledge Research in Place, Space and Politics
Edited by Professor Clive Barnett
University of Exeter, UK

This series offers a forum for original and innovative research that explores the changing geographies of political life. The series engages with a series of key debates about innovative political forms and addresses key concepts of political analysis such as scale, territory and public space. It brings into focus emerging interdisciplinary conversations about the spaces through which power is exercised, legitimized and contested. Titles within the series range from empirical investigations to theoretical engagements and authors comprise scholars working in overlapping fields including political geography, political theory, development studies, political sociology, international relations and urban politics.

For more information about this series, please visit: www.routledge.com/series/PSP

The Challenges of Democracy in the War on Terror

The Liberal State before the Advance of Terrorism

Maximiliano E. Korstanje

 Routledge
Taylor & Francis Group

LONDON AND NEW YORK

First published 2019
by Routledge
2 Park Square, Milton Park, Abingdon, Oxon OX14 4RN

and by Routledge
52 Vanderbilt Avenue, New York, NY 10017

First issued in paperback 2020

Routledge is an imprint of the Taylor & Francis Group, an informa business

British Library Cataloguing-in-Publication Data
A catalogue record for this book is available from the British Library

Library of Congress Cataloging-in-Publication Data
A catalog record has been requested for this book

ISBN 13: 978-0-367-58762-8 (pbk)
ISBN 13: 978-1-138-60932-7 (hbk)

Typeset in Times New Roman
by Sunrise Setting Ltd, Brixham, UK

Contents

Foreword

Professor Max Korstanje is a prolific scholar who concentrates on publishing terrorism research in English and Spanish (Google him to see examples). Fortunately he has put together this insightful book which explains the philosophical theories underlying terrorist behavior along with the problematic government practices that historically caused terrorism. One key difference between historical and modern terrorism is that in the past the risks were localized to specific regions (e.g., Nazism in countries bordering Germany) whereas now everyone is truly at risk of global terrorism. A glance through the world news is enough evidence of this. Therefore, many stakeholders will benefit from reading this book, which will generalize to government policy-makers around the world, strategic supply-chain management planners at private as well as nonprofit organizations, university professors along with their students, and other scholars.

If the above information has still not convinced you of the value of this book, then perhaps a few insights will help. Actually, terrorism is not a new phenomenon since, according to Korstanje, it is documented in various forms including violence, torture and poverty, throughout our history. Perhaps terrorism diminished after the Anglo-democracy era when African nations and the USA successfully fought against British colony rule, or when citizens in Argentina and Chile overthrew dictatorships. However, terrorism seems to have advanced without our noticing it, much like the boiled frog syndrome. One insight that Korstanje provides is to show us that democracy is not the freedom we once thought it was; rather we are gradually losing our human rights despite all that the developed nations have done over the years to improve equality. Korstanje describes this as a 'culture of narcissism' or 'Thana-Capitalism', and he goes on to discuss the underlying causes of global terrorism including: What do democracy, liberalism and zombie apocalypses have in common with global terrorism? Korstanje weaves an interesting account of these and how we now live with a ticking bomb. Let's hope we can leverage Korstanje's insights to disarm that global terrorism ticking bomb.

—Dr Kenneth David Strang
SUNY Plattsburgh, NY, USA

Preface

Weapons of mass destruction is now the number one global risk impact level according to the World Economic Forum (http://www.weforum.org, March 14, 2018), an international nonprofit foundation based in Switzerland. Global terrorism, including terrorist and cyber-attacks, are the third and eighth most likely global risks (World Economic Forum, March 14, 2018).

It is good research practice to triangulate the evidence so I examined different data sources. Insurance companies and business continuity providers also feel that terrorism is a significant threat. Allianz, one of the largest multinational global insurance corporations, ranks business interruption, cybercrime, terrorism and civil political events as four of the top ten risks facing businesses around the world based on more than 1900 subject matter experts from 80 countries (www.agcs.allianz.com/insights/white-papers-and-case-studies/allianz-risk-barometer-2018, March 14, 2018). Natural disasters is the other global risk that the World Economic Forum and Allianz predict, but while there are numerous climate change studies taking place, we certainly do not have enough global terrorism research.

I have personal data on the risk of global terrorism (some of this was captured in several studies that I published). Firstly, historically, my relatives served and were tortured at the hands of German Nazis. Believe me, the stories that my grandfather told me about what the Nazis did to him and knowing his medical problems (including limping and being unable to have children) left no doubt in my mind about the worst of what can happen when human rights are diminished. Secondly, most nations have increased security to combat terrorism, and here in the USA anyone arriving at a border by land, air or water basically has zero rights based on recent legislation after 9/11. Don't get me wrong – we want our country to be safe and I imagine this is true for every other nation. However, does anyone wish to end up in the interrogation room where you have no rights, no legal representation, and could theoretically be locked up forever, at the hands of the military? A colleague of mine from Australia (who is a medical doctor) sadly got caught up in this loss of human rights situation as he flew in to attend a conference when he was asked where he was actually born (he was born in the Middle East but was adopted in Australia when he was an infant). Thirdly, I was near to an actual terrorism event several years ago where the fear of terrorism was real to me and I felt helpless.

Acknowledgements

This book was a personal effort which would never have been possible without the help of many people. I thank my colleagues Rodanthi Tzanelli, Adrian Scribano, Geoffrey Skoll, Max Abrahams, Luke Howie and David Altheide for the fruitful feedback given over the years as well as my wife Maria Rosa Troncoso and my children Ben, Olivia and Ciro. I also express immense gratitude to Routledge (above all, to Ruth Anderson and Faye Leenrik) for the opportunity to publish my work.

Introduction

Numerous works have emphasized the normative explorations of democracy as well as the role of constitutionalism and representation as valid sources to protect the rights of citizens. For some reason, the literature misjudged the nature of democracy exploring only the power of suffrage, party competitions and elections, leaving many other questions open. The twenty-first century pushed political scientists to reflection beyond the paradigms of democracy, or at the least adapting the democratic institutions in the years to come. The former centuries taught a lesson, that in the name of democracy atrocious crimes can be perpetrated. The extent to which postmodern democracies will repeat the mistakes of other centuries has never been tackled (Bellamy 2008; Schmidt 2008; Stimson 2008; Canovan 2008). This is one of the main goals of this book and the fire that keeps me active as I write this project.

Historian Jane J. Mansbridge (1983) separates unitary democracy from adversary democracy. In that way, she introduces a new reading of the term, which helps to expand the current understanding of modern politics. The unitary democracy relates to concepts of unity, friendship as she puts it, in the same way that democracy was in ancient Athens. The ancient city was a place of deliberations and discussions but a real "portrait of equality" among the free citizens. In contrast, the adversary democracy rests on the opposition of interests, and the struggle of many actors to impose their own agenda. She contends, in fact, that this does not mean two types of pure democracies exist, but both coexist in the same society. In some contexts, unitary democracy surfaces while in others an adversary climate is imposed. While the nation-state, as a cultural and political project, expanded, deliberative democracy sets the pace for adversary democracy.

> The fourteenth, fifteenth, and sixteenth centuries in Europe saw a feudal, traditional, and theoretically immutable system of just prices, discovered laws, and personal ties transformed into a national, fluid, and permanently transitory system of shifting prices, positive laws, and mobile, self-interested citizenry.
>
> (Mansbridge 1983: 15)

Against this backdrop, the emerging market not only demanded further movement of labor and capital from one city to another in a quest for better opportunities for profit, but citizens were free to be contracted at any time depending on the dynamic of the market. Self-interest was situated as the milestone of the modern world, whereas a new political order, which was based on the Hobbesian terms of conflict and discrepancy, fostered a new cultural ethos. In perspective, the desire for personal achievements and individuality were prioritized over the values of friendship in ancient Greece. Mansbridge discusses critically to what extent modern democracy should be understood as the game of losers and winners instead of a world of consensus, but what is more important, the introduction of freedom as the main cultural value of society caused two important effects. On one hand, it accelerated a long-dormant mobility, placing classes into conflict. On the other hand, the free play of demand and supply evinces a sum of contrasting voices and viewpoints which are domesticated by the monopoly of law. Only the State and no other organism is capable of granting the order. Politicians pass laws according to their interests, but they should be subject to elective voting. Citizens often deposit their trust in people they do not know or with whom they have no familiar connection. The logic of laissez-faire which rules the modern economy corresponds with the vindication of self-interest, Mansbridge remarks.

> At bottom, this theory of adversary democracy is remarkably similar to modern laissez-faire economics. Following a modified version of Adam Smith´s Wealth of nations, laissez-faire economists not only accept the marketplace vision of a society based on self-interest but make it an ideal. The belief either that the invisible hand of supply and demand will aggregate millions of selfish desires into the common good, or that, because no one can know the common good, the aggregation of selfish desires is the best substitute.
>
> (Mansbridge 1983: 17)

The main thesis held by rationalists coincides with the belief that the distribution of power through the vote is the most efficient vehicle to balance the discrepancies that happen in the laissez-faire society. All citizens are equal before the law, and of course, the market, cooperating or confronting to protect their own interests. For them, this is the superior ideal from where the West was consolidated as a global project respecting other cultures. The Westernization of the world morally expressed the supremacy of Europe over other voices. In so doing, *adversary democracy* was embraced as the best of possible forms of government.

As Mansbridge tested, the ideals of political equality should not be compared to democracy but just comprehended as a means towards democratic institutions. The political equality in some contexts, when it is subject to the logic of means-and-ends may create asymmetries in the levels of powers. In some cases, the inequalities of power do not necessarily usher the society into exploitation. She cites the example of pilots and their passengers: although they all look to the same end (and have no disputes in their wish that the airplane lands safely), they perform different roles. The same relations can be observed between parents and children and there are many other examples.

The crucial step is, of course, recognizing the right of all citizens in a democracy to have their interests represented equally in a political process. In a small assembly, democracy is informal, but it should still be recognized as such. In a larger polity, the representation must become formal. In both cases, the polity gives up on the idea of providing each citizen with equal power, and allows representatives, whether self-elected or elected, more power than their constituents.

(Mansbridge 1983: 251)

Another point of discussion which deserves our attention is how the ideal of democracy resonates in non-Western countries. To take a closer look, it is important to delve into the history of Korean film to expand the current understanding of the impacts of democracy and modernity in cultural tradition. The movie industry in Korea had different stages, which ranged from the reactionary spirit against the Japanese occupation to resistance to the economic programs of neoliberalism. In the book *Korean Film*, professors Eungjun Min, Jinsook Joo & Han Ju Kwak (2003) offer an in-depth review of the plots of more than a dozen films. From its inception, Korean film developed a strong resistance to Japanese culture and the arbitrariness of an occupation which lasted from 1910 to 1945, once the Empire of Japan was defeated in WWII. This traumatic event touched the hearts of Koreans as never before, with some after-effects to date. The term *haan* denotes a type of nostalgia and suffering respecting an external oppression. In this vein, the transitional governments faced a combination of weak democracies and despotic totalitarian governments. The rise of democracy, just after the 1990s, opened the doors to a new facet of Korean cinema, which was related to the introduction of globalization and American cinema. A great portion of films, which were released from 1990 onwards, shows much deeper concerns revolving around the exploitation of the workers, at the hands of capital owners, as well as the complicity of politicians (the State) with the capital. It created a romantic reaction that woke up some sympathy for Korean traditions. The marginalization of some actors would be the inevitable consequence of some liberal policies which are oriented to make the rich, richer, and the poor, poorer. The book is not a treatise on liberalism nor does it explain the roots of capitalism, but sheds light on two important axes that guide my argumentation: the configuration of democracy – as something other than an ideological platform – as a form of government, which combined internal freedom with external oppression, has been situated as an ideal to follow. Secondly, while underdeveloped nations did not find any valid solutions or answers to the asymmetries of democracy, which are negatively amplified by globalization, more resistance or romantic expressions have emerged. These reactionary discourses allude often to a romanticized version of tradition and past, which have nothing to do with the historical facts. At the time more globalization pushes traditional cultures into modernization, more radical voices invade the minds of the frustrated city-dwellers. Not surprisingly those responsible for the terrorist blows perpetrated on Brussels, London and Nice were natives of the societies they attacked. This book centers on deciphering the intersection of liberality and terrorism, filling a gap which remains in the specialized literature.

Such a discussion begs some pertinent questions which this book will attempt to answer such as: what is a democracy? Is democracy an ideological instrument of control? What are the differences between modern and ancient democracy? What are the essential features of the (neo)liberal state? Is democracy the solution to deter the rise and advance of terrorism? Why are the so-called more democratic countries suspected of torture and human rights violations? Why do terrorists target tourists? To what extent is torture enough? And, what is the role of democracy in this new order after the stock and market crisis in 2008?

First, Norberto Bobbio (2005) understands that liberal democracy does not reach the basis of egalitarianism and equilibrium that liberals erroneously believe. He toys with the idea that there is a manifest tension between socialism and liberal democracy. The liberal democracy not only leaves some unmet promises, or people relegated from the distribution of wealth, but fails sometimes in the realization of basic rights. Ideologically, democracy claimed itself to be a better form of government in comparison with other ancient times or forms of dictatorship and monarchy. This so-called superiority is strongly associated to the ideological forces of capitalism and the efficacy of the nation-state to domesticate the citizens. In parallel, other critical voices claimed the deficiencies of contemporary democracy over recent decades. Carl Schmitt (1985) manifested the theory of exception, which was previously extracted from Donoso Cortes. In contrast to positivism and Kelsen's theory, Schmitt endorses the credibility of state and the monopoly of force to those agents who take the lead in the decision-making process. In that way, the sanction of laws is not enough to exercise power. While the ruler makes the decisions respecting some issues, it is unmarked or obliged to abide by the rules. As Schmitt overtly recognizes, there is nothing like morality in politics; even Schmitt argues that the law expresses the dialectics between oppressors and the oppressed.. Some laws may be legal but not ethical and vice versa. To put this bluntly, the powerful ruler keeps some margin to alter the laws, the legislation and change ethnicity according to its desires. Hence leadership seems to be the key factor of making politics (Schmitt 1985; Scheuerman 1999). Some more moderate thinkers like Touraine and McDonald (1994) confirm democracy should be valorized as a social construct and not a form of government. However, if there is a conceptual basis democracy should meet, it is the sum of all individual rights of the members of society. Democracy rests on three fundamental tenets: the respect for individual rights, the representation of citizenry which governs through its representants, and the limits to the ruling elite. For Robert Dahl (1989, 2005), we must distinguish ancient from modern democracy. The concept, as well as its application, three ideas of three civilizations are present in the democratic ideal: the values of equality of Ancient Greece, the division of powers of the Roman Empire, and the belief in a checks-and-balances power of Great Britain. Giovani Sartorio embraces "pluralism", an academic school within political science, to alert that there are plenty of different democracies which historically coexisted. Today, the etymology of the term reveals a type of self-government or the idea that common people may govern themselves (Sartori 1965).

Against this backdrop, Chambers & Salisbury (1960) agree that there is a clear-cut dissociation between the ideal and the real democracy, which reflects the interest and preoccupation of a whole portion of the specialized literature. They hold the hypothesis that the main limitation comes from a contemporary sensation of discrepancy between the idea and the practice, which was nourished by relativism. In view of this, there is a gap that resulted from the over-valorization of democracy – as the cure of all evils, which was a tradition inherited from Western civilization – and the need for politics to attend to the citizens' problems. During his journeys through America, Alexis de Tocqueville (2003) realized that democracy does not resolve the problems at all, but narrows the social ties into a climate of egalitarianism and mutual respect. Most likely, one of the seminal books on democracy was authored by Sheldon Wolin (1993) under the title *Politics and Vision.* Per his stance, democracy calls for a paradox because it expands the possibilities of individuals to enhance their personal achievements, but at the same time, it is very difficult for an over-bureaucratic state to address these plural claims. This quandary helps many citizens to experience their political gratification beyond what the classic parties offer. In 1993 Elisabeth Noelle-Neumann published *The Spiral of Silence* to demonstrate the correlation of advertising in politics and the influence of media in the public spheres. Per her outcomes, individuals are afraid to be excluded from what the majority thinks. Then, in elections, people are sensitive to manifest their intention to vote for the most popular candidate, while they really vote for others. Noelle-Neumann amply discusses the dissociation between what people say and finally do. The findings were continued by Weaver, McCombs & Shaw (2004) who dangle the possibilities that a candidate may be designed as a product to be exchanged through media consumption. In this respect, Christopher Lasch coins the term "the culture of Narcissism" to denote not only the lack of engagement of citizens by politics but the obsession for politicians to embellish their image instead of making concrete politics. The contemporary politics is embedded with a culture of aesthetics which alienates lay-people from the real interests of professional politics (Lasch 1991).

The 1970s was a turning point in the passage from an industrial to a service society. The term creative destruction applies not only to the thousands of jobs and plants that were dismantled but also to the creation of new creative opportunities. As a result, the economy suffered labor market flexibilization as never before, which ushered Americans into a hyper-competitive world. As Richard Sennett studied, the ideals of the industrial world which crystalize in a stable job, durable ties and appropriate conditions of labor, which were characteristic of the welfare state, set the pace for new cultural values based on individualism, the need for constant change and an extreme competition for progress and self-achievement. The industrial ideology expressed some concerns for the workers' exploitation at the factories while the new postmodern ideology gave workers the liberty to co-manage their own destiny. Beyond the so-called climate of freedom they enjoy, at the bottom lies a new system of subordination where the State slides from its responsibilities to protect citizens' rights. Those citizens who fail to innovate simply perish, in the same way as those who cannot adapt to more extreme atmospheres of competition in the labor market. This exhibits a real "corrosion of character" encouraged by neoliberalism (Sennett 2011). With a robust empirical background, Steven Davis, John Haltiwanger & Scott Schuh (1996, 1998) set forward convincing evidence that suggests the capitalist economy has cycles of crises, where job destruction outweighs job creation. Articulating different statistical bases, which are fertile grounds for economists, the experts lament that after the 1970s the US manufacturing sector suffered a slowdown of production which mutated to service-related industries. In consequence, they conclude, one of the challenges of democracies is to intervene in economies to mitigate the impacts of job destruction. Technically, these economists validate the assumptions and risks at which globalization hints.

> The labor market consequences of international trade occupy a prominent position in many discussion of economic policy. Although the effects of trade development on the level and structure of wages have received much attention, and rightly so, many commentators also express concern about the effects on unemployment levels and job security. These concerns were of paramount debate over the North American free trade Agreement.
>
> (Davis, Haltiwanger & Schuh 1996: 174)

Per their viewpoint, the openness of the US to the world engendered serious problems in the job markets. This is particularly harmful for human capital and the educational system which is exposed, if not subordinated, to the increasingly volatile labor market. The evidence implies that economic openness does not necessarily ensure a rapid decomposition of the social fabric. There are interesting studied cases where state regulation assists in the reeducation of people in new positions. Is this mobile world more uncertain than others?

In the edited book *New Risk New Welfare,* Taylor-Gooby (2004) explores the asymmetries of postmodern capitalism in fields relating to human capital, demography, production and social security. After the 1970s, the expert agrees, the number of active workers slumped because of demographic problems. The gradual decline of births, associated with an ever-increasing number of retired workers created a vicious circle in which the active workforce was not enough to support the costs of retirees. This problem was aggravated by the introduction of technology and robots in the factories, which notably reduced job opportunities for the middle and lower classes. The main thesis here is that the more expansive the capitalist economy, the fewer jobs and higher taxes for the working classes. Zygmunt Bauman (2001) puts emphasis on the commoditization of the workers, who are often sold as products at the labor marketplace. This means that consumers are transformed into slaves of the goods they often choose. The classic definition of the market as the encounter between the demand and the offering was replaced by new connotations, where there are no clear-cut borders between the processes of production and consumption. Bauman uses the term "fetishism of humanity" to denote the remains derived from the system of production. While many free citizens are left behind the opportunities that the consuming world offers, it is no less true that consumerism was established as a new type of social covenant which governs the life of citizens. Elaborated to be annually replaced, goods in these liquid times are not durable, in the same ways that social bondage is fragmented. Bauman's insights reveal two important aspects of modern democracies. Firstly, politics are subordinated to the organizations of economies as well as the means of production. Secondly and most important, social behavior is previously determined by these material means of production disposing and manipulating emotions towards an emptied-imagination (Bauman 2001).

Quite aside from this, the polemic is still open to date. This book is not a much deeper reflection that seeks to grasp a consensual definition of democracy. Rather it borders the dichotomies, limitations and problems that contemporary democracy has to face in dealing with terrorism.

How this book should be read

Though chapters may be read separately, all share a common thread. Western civilization has developed a closed hospitality, which regulates not only the citizens' lifestyles but the position of "non-Western" Others as outsiders. Doubtless, one of the myths that embodies the mistrust of the West for aliens seems to be the Trojan Horse, which serves as a metaphor for the risks hospitality supposes. The 9/11 attack accelerated the outbreaks of racism and Islamophobia which were encapsulated in the core of liberal democracies (Korstanje 2017). Hence, the first chapter explores the political nature of liberalism and the liberal state. Lay-citizens are often educated to think consuming, delivering children to school and paying taxes is the best way of exercising democracy. In view of that, the liberal state posits a bipolar logic, we and they, in which case "the alterity" is seen in sharp opposition to the current cultural values of the West. If Westerners believe they are smart, educated, civilized and democratic, the aliens turn undemocratic, prone to despotic governments, lazy and so forth. This represents a big problem at the time of defining the ethical borders of security. From its inception, in the ink of Thomas Hobbes, John Locke and other liberal writers, the liberal state symbolically and culturally monopolized legal violence to keep an internal and durable peace. The doctrine of sovereignty and security occupied a central position in the configuration of nationhood and the modern state. Besides, the over-valorization of democracy paved the way for the decline of critical thought, which worked in favor of the rise and expansion of capitalism. As one of the most consolidated democracies, in theory, the US pointed out that the lack of stronger institutions was the key reason for the upsurge of terrorism. This was exactly the reality lived in Latin America during the 1970s. While juntas and Pinochet orchestrated a repressive program to eradicate communism, the US adduced the human rights violations as the lack of a democratic tradition in the southern hemisphere. Accustomed to living through a set of different coups, Argentinians and Chileans failed to improve the checks-and-balances power to avoid dictatorships. The application of torture, as well as other horrendous crimes, was the direct result of the historic indifference these cultures have for democracies. Of course, Latin America learned the lesson, but the US was never immune to the authoritarian spirit. The main point of entry in this discussion is that terrorism obliges a rethink of the nature and performance of democratic institutions as well as the role of politicians.

The second chapter traces back the history of consumption and the socio-cultural conditions that influenced the passing of a society of producers to a society of consumers. We put in dialogue two senior scholars, Zygmunt Bauman and Kathleen Donohue to expand the current understanding of capitalism and its connection with mass-consumerism. While Bauman posits that the society of consumption created an extreme competitive culture where citizens are commoditized, Donohue explains that Marxism and its exaggerated fears of the advance of poverty were the real reasons behind a substantial shift in the classic economic theory. Without exception, from time immemorial economists have maintained that uncontrolled consumption was defective for the well-functioning of the economy. However, after the 1970s, a bunch of left-wing writers emphasized the need to adopt consumption as a valid alternative for poverty relief. Donohue confronts Marxism adducing a more than interesting hypothesis. Capitalism engendered a "freedom from want", which was re-channeled towards the liberalization of ties, but instead of puritanism being responsible for this, Marxism involuntarily reinforced the ideological discourses of liberalism.

The third chapter continues the legacy left by Jean Baudrillard and Guy Debord regarding the society of the spectacle. It ignites a heated debate revolving around the "terror camps" geographically located in Israel and the West Bank, Palestine. These camps offer a morbid spectacle where tourists may emulate Israeli military officers. Far from reversing or solving the asymmetries created by global capitalism, this sinister consumption recycles and reinforces the logic of exploitation. This engendered a morbid culture, we dubbed "Thana-Capitalism", which disposes of disasters, sites of mass death, genocides and morbidity as a criterion of attraction for the vast global public. I must clarify – through this chapter – some open questions duly highlighted by R. Tzanelli in her professional book review of my work, *The Rise of Thana Capitalism and Tourism* (Korstanje 2016). The term Thana Capitalism signals an ever-increasing trend which is shifting not only the content of media entertainment but also the social institutions. Answering Tzanelli's critiques point by point, I argue that visitors, far from feeling empathy with victims, adopt a narcissistic character that neglects the Other's pain, or at best employs it as a form of pleasure-maximization. Lay-citizens, in Thana Capitalism, feel special and gazing at how others die is a valid way – for them – to conserve an auratic status.

The fourth chapter, in consonance with the above-mentioned comments, centers on the philosophical quandary of torture. After 9/11, the most consolidated democracies have engaged in a debate respecting the efficacy of torture. Under the auspices of the "ticking time bomb theory", which suggests that in some conditions torture may anticipate the next terrorist attack, saving lives, some scholars (like Michael Ignatieff) defended a legal use of torture. However, this argument rests on some contradictions. In many ways, politicians manipulate elections, changing the law, agreeing secret consensus, or simply using some legal tricks to avoid ethnic minorities taking power. In fact, as discussed in the first chapter, the state of exception the law endorses to the ruling elite not only discourages any type of torture of those suspected of terrorism but also as we found, torture is not enough to get vital information to prevent terrorism.

The fifth chapter deals with the apocalypse theory, the world of zombies and the use of technology as the architect of modern rationality. A degraded view of the world, adjoined to anxieties about the rise of external risks, resulted in a scatological mythology, rooted in the contradictions of technology. In this respect, while technology helps Western civilization to expand and configure a vast, far-flung empire, it becomes the trap of humanity. The bottom-days theory, as well as the apocalypse, remind us not only of the vulnerability of humans in coping with nature but also the risks of greed which is often castigated by the gods. Paradoxically, a technology which was designed for making this world a safer place degenerated towards a state of disaster and emergency that pushed mankind to its marginality.

The sixth chapter, rather, analyzes the recently perpetrated terrorist attack in New York City where five Argentinians lost their lives. This chapter interrogates the anthropological nature of hospitality and tourism with a strong focus on the guest–host relationship. The attacks on NYC not only opened the doors for political instability within the US, but the impacts also resonated in Buenos Aires, where President Mauricio Macri touched on the theme in various speeches. The chapter reveals the troubling ties between terrorism and tourism, as a deep-seated issue which deserves to be investigated in future approaches.

Last, the seventh chapter connects directly with the first. It dissects the ideological core of European ethnocentrism as well as the limitations of conditioned hospitality as fabricated by the West. Here I discuss the problems and limitations of Islamophobia as a new type of neo-racism terrorism. Diverse voices have claimed that the media depicts and imposes an image of Muslim communities as potentially dangerous for the European lifestyle. It is unfortunate that the attention given to Islamophobia does not come from Anglo-Saxon researchers but is limited to Muslim scholars. Hence, we pit the contributions of Samuel Huntington (1993) (regarding the "clash of civilization") with Edward Said's (2016) seminal text, *Orientalism.* Our main thesis is that under the metaphor of the medical gaze, which looks and extirpates the infected organs (pathology) to save the life of patients, Muslim communities run a serious risk of being vulnered and ghettoized (if not decimated) by the liberal state. This is perhaps one of the paradoxes of the liberal states and neoliberal democracy.

References

Bauman, Z. (2001). Consuming life. *Journal of Consumer Culture, 1*(1), 9–29.

Bellamy, R. (2008). "The Challenges of European Union". In *The Oxford Handbook of Political Theory*, J. Dryzek, B. Honig, & A. Phillips (eds). Oxford, Oxford University Press, 245–261.

Bobbio, N. (2005). *Liberalism and Democracy* (Vol. 4). London, Verso.

Canovan, M. (2008). "The People". In *The Oxford Handbook of Political Theory*, J. Dryzek, B. Honig, & A. Phillips (eds). Oxford, Oxford University Press, 349–362.

Chambers, W. N., & Salisbury, R. H. (Eds.). (1960). *Democracy Today: Problems and Prospects*. New York, Collier Books.

Dahl, R. A. (1989). *Democracy and Its Critics*. New Haven, Yale University Press.

Dahl, R. A. (2005). *Who Governs?: Democracy and Power in an American City*. New Haven, Yale University Press.

Davis, S. J., Haltiwanger, J., & Schuh, S. (1996). Small business and job creation: Dissecting the myth and reassessing the facts. *Small Business Economics*, *8*(4), 297–315.

Davis, S. J., Haltiwanger, J. C., & Schuh, S. (1998). *Job Creation and Destruction*. Cambridge, MIT Press Books.

De Tocqueville, A. (2003). *Democracy in America* (Vol. 10). Washington, DC, Regnery Publishing.

Huntington, S. P. (1993). The clash of civilizations? *Foreign Affairs*, 22–49.

Korstanje, M. E. (2016). *The Rise of Thana Capitalism and Tourism*. Abingdon, Routledge.

Korstanje, M. E. (2017). *Terrorism, Tourism and the End of Hospitality in the West*. New York, Springer Nature.

Lasch, C. (1991). *The Culture of Narcissism: American Life in an Age of Diminishing Expectations*. New York, WW Norton & Company.

Mansbridge, J. J. (1983). *Beyond Adversary Democracy*. Chicago, University of Chicago Press.

Min, E., Joo, J., & Kwak, H. J. (2003). *Korean Film: History, Resistance, and Democratic Imagination*. Westport, Praeger.

Noelle-Neumann, E. (1993). *The Spiral of Silence: Public Opinion, Our Social Skin*. Chicago, University of Chicago Press.

Said, E. W. (2016). *Orientalism: Western Conceptions of the Orient*. London, Penguin UK.

Sartori, G. (1965). *Democratic Theory* (Vol. 579). Westport, Praeger.

Scheuerman, W. E. (1999). *Carl Schmitt: The End of Law*. Rowman & Littlefield.

Schmidt, R. (2008). "In the Beginning All the World Was America: American Exceptionalism in New Contexts". In *The Oxford Handbook of Political Theory*, J. Dryzek, B. Honig, & A. Phillips (eds). Oxford, Oxford University Press, 281–296.

Schmitt, C. (1985). *Political Theology: Four Chapters on the Concept of Sovereignty*. Chicago, University of Chicago Press.

Sennett, R. (2011). *The Corrosion of Character: The Personal Consequences of Work in the New Capitalism*. New York, WW Norton & Company.

Stimson, S. (2008). "Constitutionalism and the Role of Law". In *The Oxford Handbook of Political Theory*, J. Dryzek, B. Honig, & A. Phillips (eds). Oxford, Oxford University Press, 317–332.

Taylor-Gooby, P. (Ed.). (2004). *New Risks, New Welfare: The Transformation of the European Welfare State*. Oxford, Oxford University Press.

Touraine, A., & McDonald, K. (1994). Democracy. *Thesis Eleven*, *38*(1), 1–15.

Weaver, D., McCombs, M., & Shaw, D. L. (2004). "Agenda-Setting Research: Issues, Attributes, and Influences". In *Handbook of Political Communication Research*, L. Lee Kaid (ed.). Mahwah, Lawrence Erlbaum Associates, 257–282.

Wolin, S. S. (1993). Democracy, difference, and re-cognition. *Political Theory*, *21*(3), 464–483.

1 The liberal state

Introduction

Over the years, philosophers, political thinkers and academicians theorized on the factor that keeps society united. This, to some extent, triggered a hot debate revolving around the gregarious or individualist nature of man. In this introductory chapter, I shall discuss critically to what extent the liberal state needs the idea of democracy to survive; at the same time I shall lay the foundations for a new theory that helps readers understand contemporary politics. While lay-citizens are systematically educated to think of democracy as a positive form of government that often respects the individuality of minorities, avoiding the rise of dictatorships, terrorism and terrorists are imagined as maniacs, hatred-filled enemies of liberty and the values of respect that characterize Western civilization. Here some interesting questions arise: Are democracy and terrorism two sides of the same coin? Is democracy the direct result of modern capitalism or simply its precondition?

Toby James, professor at the University of East Anglia, poses an interesting model to expand the current understanding of democracy and its intersection with the welfare state. According to his viewpoint, though in the democratic life the power of government is limited to the interplay of actors and rules, it is no less true that the framing of such rules is politically manipulated by the elite. It is possible that democracy is considered one of the best-known forms of government, but this does not mean that the rules of the game cannot be changed according to the interests of lurking elite statecraft (James 2012).

The liberal state

The concept of the liberal state sparked serious controversies and discussion in the social sciences. Over the years, sociology found that we are accustomed to imagining the local environment as it always existed even though it was recently introduced. Through democracy, the liberal doctrine says, institutions weave the necessary political stability for the economy to be strengthened. The produced wealth would be fairly distributed to all classes and groups according to the democratic rule. In this vein, R. L. Heilbroner (2011) dates the origins of capitalism back to the conquest of the Americas. Nation-states, though they are considered as universal entities, were no older than three centuries. The original discovery and the conquest as it was articulated years later resulted not only from the technological breakthrough which ignited a military force that consolidated a vast, far-flung empire but also by the needs of introducing discovery as a new category, which marks the centerpiece of a new age that definitively abandons the Middle Age (Heilbroner 2011). This point was brilliantly assessed by Edward Said (1979) in his book *Orientalism*, where the author deploys several literary resources to describe how Western civilization has unilaterally created a biased image of the East in order to validate previously designed stereotypes and racial prejudices. Said contends that the Far East was culturally invented as a place of mystery, fraught with political upheaval and inferiors, which doubtless deserved to be conquered, civilized and subordinated to European rule. However, the Orient is something other than a simple creation; it corresponds with a dense net of narratives, sentiments and emotional dispositions which were historically oriented to legitimate the presence of England and France in Asia. Said takes his cue from the notion of discourse as earlier formulated by Michel Foucault. As a system of knowledge orchestrated to support the expansion of European powers, Orientalism acted not only to induce an ideological message, where the European identity was superior and ideal regarding another identity, but also as an exclusionary force that was legalized by modern science. The non-Western "Other" was thought of as uncivilized, less smart, or simply less virtuous than Europeans simply because there was substantial scientific evidence in the hands of scholarship. As Said puts it, in any case, Orientalism was a science, though it was consolidated as a serious academic discipline while England and above all France consolidated their military occupations in different Asian countries.

According to the previous argument, what Said and Heilbroner valorize is that the economic maturation of Europe, adjoined to a cultural trend to invent "a non-Western Other", played a vital role in the expansion and arrival of capitalism. In a seminal book, *Unspeakable Violence,* Nicole Guidotti-Hernández (2011) calls attention to the function of a discursive violence, which is intended to colonize the minds of locals. The history of colonialization seems to be today a great spectacle for visitors and tourists while the violence exerted against the local agent is hidden. This suggests that nation-states in the Americas remapped the social borders by the disposition of a formal violence that was supported by "an unspeakable violence". The unspeakable violence, which is enrooted in the language, was historically used and manipulated by the different states to subordinate some ethnic minorities.. Based on a much deeper process of differentiation – respecting other states – and the subsequent economic means of production that organized the territory, nation-states administered racism and sexism to control their citizens. Guidotti-Hernández coined the term "selective memory" to describe how history was carefully manipulated to protect the interests of the ruling elite. After all, "Race" is a concept in which elites play a key role in its construction and negotiation. Racial mixture often is used as the basis of disenabling the emancipation of ethnicities. This belief runs the risk of presenting the Mestizo or Chicano as part of nature when really they are legacies of a colonial order. In view of this, any movement of resistance is remapped and reconfigured according to new, more acceptable values rooted in the culture of the masters. In Mexico, to set an example, one could experience certain nostalgia for those aborigines who had lost their lands, but what the aboriginal evokes remains a concept politically determined by white power. The center of hegemony, like ideology, works by the control of what it means to be an 'authentic' Indian, Chicano or Mestizo. To varying degrees, scholars and intellectuals have historically contributed to this system of labeling.

In a nutshell, the process of colonialization opened the doors for unleashing the "dogs of war", in which case Europe maximized its economic resources for the creation of a war-machine ready to colonize other territories, while a paternalist discourse to preserve this non-Western Other was unilaterally imposed. From its outset, ethnology and anthropology were two academic disciplines heavily concerned by the disappearance of aboriginal cultures in the hands of modernization and European lifestyle. Fieldworkers entered tribal villages with the end of collating as much information as they could but, what is more important, to pick up artifacts, pots and clothes to be showcased in a museum. Paradoxically, the altruistic spirit of anthropologists produced a considerable framework of knowledge, which was employed by colonial officers to learn more about Indians and their tactics of war. To some extent, anthropology and colonialism were inextricably intertwined (Lewis 1973; Harris 2001; Pels 2008). As Mervin Harris (2001) observed, the rise of anthropological theory was not only imbued within the Darwinian logic but emerged as the need to reach discoverable lawful principles. The first ethnologists enthusiastically embraced the power of history as well as the national character in the formation of individual reasoning. At this stage, the notion of culture was associated to the sense of biological heredity laying the foundations for the emergence of "free will" as the sign of "natural law", which means the law shared by all humans. The rational Europeans who often make their decisions centered on their capacity to exercise free will were pitted against the archetype of the noble savage, an allegory unfolded to present a derogatory image of the natives as lazy, irrational and, even under some conditions, compared to children. In dispute with the notion of predictability, aboriginals should be educated and duly cultivated under the auspices of Europeanness (Harris 2001). This does not mean that anthropology and ethnology were allies of the colonization process and the cruelty of colonial officers.

One of the authoritative voices in the study of American history, Bernard Bailyn (2017), explores the ideological roots of the US and capitalism. From its outset, the US debated between two contrasting tendencies: democracy and loyalties to authorities. One of the aspects of British colonialism was the duality between a liberal center, situated in London, and the repression of army forces in the colonies. Bailyn pivots in deciphering the historical hegemony of the British Empire today emulated by the US in combining an internal peace which flourished thanks to the contributions of liberal thinking, and an ongoing state of conflict abroad when interventions in autonomous nations seriously confront the ideals of democracy as globally proclaimed. From its inception, American nativism developed an uncanny aversion to immigrants and the Catholic faith. As Tyler Anbinder (1992) hinted, nativism – which derives in part from the "Know Nothing" movement – reacted negatively to the would-be totalitarian spirit of Rome and the Catholic Church while they adopted an anti-slavery, anti-alcohol and anti-exploitation dogma. From its outset, nativism introduced an ideological discourse which aimed to claim that Catholics wanted to control America through the articulation of different tricks.

> Know nothing believed that the new Catholic assertiveness carried grave implications for the future of American Protestantism. With Catholic immigrants pouring into the country, they would eventually outnumber Protestants. Once they gained hegemony, nativism reasoned, their purported intolerance would lead them to ban Protestantism altogether.
>
> (Anbinder 1992: 112)

Unlike the racist South, which fostered the exploitation of slaves as its main source of production, this movement originally flourished in the North and was playing a vital role in not only encouraging future anti-slavery campaigns but also in the liberal tradition within the Republican and Democrat parties. Liberality was the main cultural value of America as well as the hallmark nation that welcomed European migrants. Europe was in moral decline and migration would wreak havoc on the free spirit of the US, Anbinder concludes. In both texts, Anbinder and Bailyn agree that the concept of liberality in America confronted the old continent. While paradoxically the fear of tyranny and intolerance led towards regimes of oppression which placed democracy in jeopardy. The lesson learned by American history seems to be that the demonization of alterity occupied a central role in the configuration of the country. The fear of Catholics was soon replaced by the red-scare, and today the war on terror. J. Simon sets forward an interesting model that explains how the obsession for the founding parents left a checks-and-balances power. The obsession for cultivating a democratic life ushered politics into the dark side of indifference, in which case, the political power envisaged the possibility to employ fear as an instrument to overcome the institutional setbacks that impede social change. The act of *governing through crime* laid the foundation for a new jurisprudence which was prone to authoritarian – if not unconstitutional – policies that vulnerated the rights of some ethnic minorities (Simon 2007).

Lastly, I have drawn a theoretical model to understand the role of Westernization in the configuration of nationhood and nation-state. To expand the current understanding regarding colonialism, as a European phenomenon, it is vital to discuss first the economic means of production that led Europe to seek to subordinate other economies. Colonialism as a cultural project needed an inferior "Other" to revitalize the economic background that led very well towards a center–periphery dependency. Needless to say, colonialism created a specific epistemology to cover the real interests of main powers. From Hobbes to Locke, many philosophers have interrogated the nature of society. Over time, philosophers deposited their efforts in the figure of dialectics to explain the world: the law-making process. However, the concept of law-making, far from being subjective, was based on the needs of inventing a third-object that can control the force. This same sense of violence which was created through the articulation of law allowed the formation of frontiers. The liberal state was torn between a process of industrialization which engendered further slums and ghettos while at the same time, the notion of security erected a barrier between the city and its outside. Equally important, since capitalism evolved according to an epistemological division as we (the civilized men) and they (the noble savages), Europe alluded to "the sacred-law of hospitality" to discipline the alterity. From its inception, hospitality was an intertribal rite oriented to self-defense and exchange of people, but once adopted by imperial powers it was defined as inherently belonging to natural law. Those tribes that did not honor hospitality were systematically defeated and dominated. Although hospitality played a crucial role to fabricate a durable paternalistic viewpoint revolving around indigeneity, racism kept the coexisting lower social classes and ethnicities under the hegemony of the white class. It was unfortunate that the alterity was understood in a one-sided way that facilitated the rise of modern terrorism globally over recent decades. It is safe to say that hospitality was historically the touchstone of Western civilization. Today Western nations are seriously affected by radicalized groups such as ISIS, which look to create political instability through the articulation of violence in tourist destinations. Likewise, the Occident has witnessed the gradual end of hospitality (Korstanje 2017).

The United States and capitalism

It is clear that capitalism originally flourished in England but was rapidly moved to the United States. One of the fathers of French sociology, Emile Durkheim, was worried not only by the advance of industrialism but also by its aftermath for social bondage. Durkheim (2014) pays particular attention to the means of production – in consonance with Marx – and the legal division of labor, in which case he innovates in a newly emerging field of sociology. After all, production involves not only different stages and agents but also finer differentiations in the task and activities. He divides the industrial world from the medieval days. Unlike other economic systems, industrialism escaped from the individual craftsman, moving towards more sophisticated activities that required more specialized professions. This is exactly the role played by education in the industrial system, educating and training the future workforce for their new roles. With a closer look, Durkheim's insight seems not to be very different than other utilitarian postures such as Bentham or Smith (e.g., *The Wealth of Nations*) unless by the fact Durkheim ignites a debate on the growing climate of alienation that lay-citizens face daily in urban contexts. To some extent, Durkheim's worries vaticinated that social bonds would be seriously harmed by the advance of industrial rule. Through the articulation of mechanic and organic solidarity as a guiding model, he holds the thesis that human evolution was prone to mark the passage (rise or decline) of society towards a new stage. As a founding event, *industrialism* undoubtedly implied the definitive abandonment of the Middle Age and the agrarian economy. While mechanical solidarity ignites a process of integration which is based on tradition and homogeneity, organic solidarity refers to the specialization of the roles that sophisticated societies need. The touchstone of Western civilization is the contract, which endorses legal duties between two or more parts. Contracts are gradually replacing old forms of solidarities leading society towards an irreversible climate of instrumentalization and of course depersonalization. At the same time, while societies which are characterized by a mechanical solidarity are based on penal law, organic ones organize life around "restitutory law". Although, Durkheim acknowledged, increased production involved a major degree of specialization, which brought social ills such as anomie, suicide, individualism and addictions (Durkheim 2014).

As the previous argument showed, Durkheim in this sense can be placed in dialogue with Max Weber (although in fact they never exchanged an epistolary correspondence). Unlike Durkheim, who is claiming for British and French industrialism, Weber turned his attention to daily life in America. It is important not to lose sight that part of his observations come from personal travels to the United States. However, in sharp contrast to Durkheim, Weber (2002) eludes positivism or functionalism as the centerpiece of his explanations. Rather, he reminds us of the influence of Puritan faith in the configuration of capitalism as a cultural project. Weber discusses critically the roots of modern capitalism as the result of religion decline. To a greater or lesser degree, Protestantism gradually introduced a new obsession of material accumulation (in the US) as a vital force that cataloged the doomed souls from the saved ones. In perspective, the sense of predestination contains the names of those who would be saved or those whose entrance to heaven would be granted. Preliminarily, Weber discovers how Protestant workers developed a more intense attachment to work than Catholics. He argues convincingly the problem following this axiom: that over-production and amassed wealth serve to show Protestant workers that they are part of the "chosen peoples", the most virtuous in the eyes of the Lord. This is the reason behind the success of Protestant nations to situate at the top of mature economies in the capitalist system, and at best for Weber, the intersection between capitalism and religion. Let us clarify for readers that Weber knows, for example, that capitalism's origins are manifold though, as he insisted, the religious aspect is of paramount importance towards its consolidation. This still is in consonance to what Patricia Nelson Limerick dubbed "the unbroken past with American West". Far from the stereotypes and prejudices in popular parlance which led us to think that the history of the US is the eternal struggle between the industrialized North and ever-agrarian South, the West was of significance in the crystallization of an American character, which was fundamentally necessary to the rise of capitalism. The passion for work, as well as greed, real estate and of course speculation, were mainstream cultural values provided by Westerners to forge capitalism. In view of this, as Limerick amply shows, the conquest of the West not only was not easy but sometimes rested on speculation and wrong information. The colonizers often victimized themselves when they were involved in attacks or conflicts with neighboring Indians, especially when they intruded on other lands. Needless to say, religion and their faith played a crucial role in mitigating the effects of daily frustrations but more important, by making the New World a safer place to live. Through the introduction of property rights and the derived configurations of power around owners and ownership, the US started a new tradition that notably distanced it from England: *American capitalism.* As Limerick puts it,

> In finishing what God had begun, transforming the desert into gardens, Americans would usher in a new era of human partnership with God. And, in a happy overlap with America's other social problem, the displaced native white middle class, the surplus men, could find their opportunity in the desert West. It would be a new frontier both in geographical expansion and an opportunity for those penned up and in need of a refuge.
>
> (Limerick 1987: 136)

One thing seems to be important: the notion of culture is a social construction no older than the nineteenth century, originally created to mark a difference with the alterity. The spirit of Europe understood the world in black and white terms; while civilization represented the progress and the ideals of enlightenment which was uniquely cultivated in Europe, savagery was esteemed as a state of childhood subordinated to the rationale. Although the idea of democracy exerted some control over government, some ethnic minorities such as black people, Indians and recently even Latin Americans have been systematically relegated (Martin 1964; Limerick 1987; Baker 1998; Skoll & Korstanje 2013; Korstanje 2017). Once again, Americans appealed to the archetypes of liberalism and democracy as their ancestors (Britons). On the surface, some European ethnicities, who arrived in the US after the 1980s, were amalgamated with Anglo-Saxons while others were repelled.

Basic rights

Through the 1970s, Mancur Olson (1971) set forward an innovative model to understand the functioning of market and nation-state, in what he dubbed as the theory of collective goods. He starts from the premise that individuals preferably move in protecting their own interests, instead of helping others or pursuing the in-group goals. Olson was strongly interested in developing a conceptual framework that explains holistically the evolution of pressure groups. The collective goods should be understood as the achievement of common goals as well as the satisfaction of shared interests, which gives members an incentive to join. As Olson contended, one of the problems of the market seems to be related to the costs of entrances. Each firm looks to its own profits, entering into competition with others firms. While each organization certainly moves individually to gain further segments, it ushers the firms into a war of prices, which sooner or later jeopardizes the functioning of the market. In order that this does not happen, the theory of collective goods says that the State should regulate (alternating constraints and incentives) public behavior. Otherwise, agents rationally look at the maximization of benefits ignoring the system as a whole.

There is no other author who has explored the effects of liberalism in politics as has John Rawls. In his work, *The Law of Peoples*, he holds that any government is ethically forced to intervene with others for assistance or in quest of any interest. His theory reasonably orients to explain how justice works according to five forms of political organizations. Let us start with what he understands as the "laws of peoples". Preliminarily, the term comes from a utopia imagined as a world where all nations embrace democratic government as their form of government. One of the chief goals of Rawls is providing an in-depth description that helps us to understand why some nations fail to organize their life into liberal or democratic values (Rawls 1999). To put this in other terms, why do modern democracies not fit all cultures and nations?

To answer the above-formulated question, he adds, one of the evils of human history seems to be not only unjust war but political oppression that internally leads to deep-seated injustices. As the previous argument shows, five subtypes of political organizations may be framed into Rawl's development: a) reasonable law, b) decent people, c) outlaw-states, d) societies burdened by unfavorable conditions and e) benevolent absolutism.

The theory of reasonable law rests on the belief that people sacrifice their appetite for war and ambitions if wider forms of political, economic and social cooperation are achieved. Therefore, trade and negotiations are of paramount importance to balance international relationships. Rawls sets forward a model to understand how the conflict may be undermined. The veil of ignorance not only facilitates an egalitarian dialogue among liberal peoples but also determines the lack of self-determination. People behave without knowing their own possibilities or the settings where they move (Rawls 1999). Similarly to other liberal thinkers, Rawls contemplates that people are free and independent by nature. Their bonds are part of a much broader covenant where self-defense is based on the respect of rights. In this ideal world, Rawls envisages that democratic countries appeal to peace to construct a durable state of cooperation among states (Rawls 1999).

Needless to say, some voices have exerted a radical criticism of the role of democracy and liberalism which are politically distorted to legitimate the material asymmetries of capitalism (Raventós 2008; Murillo 2008). Robert Sugden discusses critically the extent to which democracy is based on a complex interplay of an economy of rights, which leads to cooperation or conflicts depending on the needs of maximizing gains by the minimization of costs.

> Individuals living together in a state of anarchy, I shall argue, tend to evolve conventions or codes of conduct that reduce the extent of impersonal conflict: this is spontaneous order. The origin of these conventions is in the interests that each individual has in living his own conventions without coming into

conflict with others. But such a convention can become a basic component of our sense of morality.

(Sugden 2005: 8)

While democratic order deals with two types of conventions, which are enrooted in the natural law, the conventions of cooperation and property, the society stays united through the articulation of reciprocity. As a zero-sum game, the interest of some players to cooperate in adversity contrasts notably with the needs of protecting ownership before a third party. The sense of reciprocity, Sugden adheres, gives the necessary stability to the system in order to ensure fair play for citizens, and of course, invoking the intervention of the State when some of the basic rights are vulnerated. In contrast to liberals, this argument suggests that democracy or dictatorship are different forms of organization, which relate to an economic preexisting order.

In this respect, thinking of democracy as the best of all worlds seems to rest on shaky foundations as B. Susser brilliantly stressed. Political governments often come across a bipolar logic which alternates dictatorships with populist democracies. In sum, political systems are enrooted in cultural environments that not only precede the daily behavior but also the ways of making politics. Every society engenders a proper climate of politics which, depending on the variables involved, would develop totalitarian or democratic regimes (Susser 1992). As Dahl (1992) puts it, one of the main limitations of political scientists consists in overlooking the history of nations, as well as the character to embrace democracy or dictatorship according to their contextual interests. The desire to see liberalism in all countries was not only a prejudice, which was enrooted in the colonial order, but also an ideological discourse of colonialization employed by the UK earlier and the US today. D. Easton, a philosopher of politics, argues that politics as a science emerged as an obstacle, an impediment that circumscribes the citizen into ignorance. Since lay people are not familiar with the ethical boundaries of politics, because they are far away from it, the study of political understanding is a way of controlling the performance of the State (Easton 1992). From its inception, Easton adds, political science has debated the dichotomy of democracy and efficiency. While the former grants egalitarian access to justice for all, the latter legitimates the authority of governors. Almond & Verba (1992) confirm that democracy and dictatorship have their differences. One of them seems to be the possibility of administrating their own decisions. Democracy looks to people to participate in public space whereas totalitarianism evokes the submission to the stronger. Nonetheless, under some circumstances, a democracy may become a dictatorship. Furthermore, C. Bay acknowledges that politics surges from the diversity of views, which take education as a form of socialization to achieve good coordination. But at moments when politics set the pace for pseudo-politics, which means that personal power will blur the ethical boundaries of diversity, society erodes its democratic basis (Bay 1992). Senior economists Haggard & Kaufman have revealed how coups and democracies coexist in the threshold of time. Basing their case on the analysis of almost twenty case studies that involve the four continents, the authors suggest that legitimacy and how it is managed by the elite is the key factor that explains the transition from democratic to non-democratic government. To some extent, the failure of economic policies harms social cohesion, in which case the political system appeals to coups in order for the system not to be fragmented. For Haggard and Kaufman, far from being stable or closed on its own, democracy mutates towards other forms according to the efficacy of the ruling elite to manage the economy (Haggard & Kaufman 1995). It is safe to say that democracy allows the exchange of some rights (to protection, consumption and of course the right to strike) that some dictatorships neglect. Liberalism and capitalism are inextricably intertwined because one cannot exist without the other (Donohue 2003). However, this moot point raises another open question, why are some rights more significant than others?

In the seminal book *Basic Rights,* Henry Shue (1996) interrogates the essential nature of rights. As he puts it, basic rights are fundamental simply in view of the fact that, without them, other secondary rights cannot be achieved.

> Security and subsistence are basic rights, then, because of the roles they play in both the enjoyment and the protection of all other rights. Other rights could not be enjoyed in the absence of security or subsistence, even if the other rights were somehow miraculously protected in such a situation. And other rights could in any case not be protected if security or subsistence could credibly be threatened.
>
> (Shue 1996: 30)

This argument fits very well with the idea that there are no primary or secondary rights or positive and negative rights, unless by the action of the protective institutions which are oriented to protect citizens. Basically, as Shue acknowledges, popular opinion judges positive rights as those where the person engages with a coactive attitude while negative rights contrarily involves a passive situation where the person retreats. This hypothesis, for Shue, not only rests on shaky foundations but is false. Neither security produces negative rights nor subsistence positive ones, simply because in order for security to be enjoyed by citizens, others such as officials, politicians and the executive branch should adopt a proactive role. The opposite is equally true: economic subsistence needs cooperation to produce "food", "houses" or anything else; in any case, Shue holds the thesis that many citizens lack the substance of their subsistence rights which are paradoxically caused by external threats created or mitigated by others. This is particularly true in underdeveloped nations where a great portion of the population cannot meet their basic needs (Shue 1996). Shue's contributions shed light on the conceptual limitations of capitalism to operate in the horizons of exploitation and scarcity. In fact, whenever all rights are successfully met there would be no further limitations to speak of "basic rights", but what does happen if citizens move in a platform of scarcity? While the rights of subsistence are previously determined by others' actions, it is safe to say that my behavior would affect others in their ontological security. After all, capitalism rests on a climate of competence where the own quest of business seems to be articulated in what others may potentially lose. As Shue eloquently writes,

> It is extremely difficult merely to mind one's own business amidst a scarcity of vital commodities. It is illusory to think that this first commandment of liberalism can always be obeyed. The very scarcity draws people into contact with each other, destroys almost all area for individual manoeuvre, and forces people to elbow each other in order to move forward. The tragedy of scarcity, beyond the deprivations necessitated by the scarcity itself, is that scarcity tends to make each one's gain someone else's loss.
>
> (Shue 1996: 46)

To cut a long story short, liberal capitalism as a cultural project, which transcends the economic order, imposes a logic of dispossession which is temporarily sustained by the trends toward *creative destruction.*

Democracy and the dilemma of torture: when democracy is not enough

September 11 of 2001 and the bloody attacks perpetrated against the Pentagon and the World Trade Center changed not only the rules of international relations but the ways non-Western "Others" were depicted. David Altheide (2017) eloquently observes that the aftermath of 9/11 still affects the existent legislation, especially in migratory matters, paving the way for the rise of radicalized discourses. Donald Trump reached the US presidency after promising to build a wall to stop illegal migration coming from Mexico. This raises a more than pertinent question: is democracy at risk?

In his book, which is entitled *Counter-Terrorism and the Prospects of Human Rights*, Professor Ipek Demirsu (2017) argues that contemporary society is based on the confrontation between two different needs: security and democracy. While the former gradually escalates into a popular climate where citizens often ask for further controls over migrants and surveillance technology at borderlands, the latter refers to the constitutional liberties the government should ensure as the precondition towards a peaceful coexistence. The English-speaking countries unilaterally imposed a process of securitization which is adopted in many underdeveloped nations. Here a paradoxical situation surfaces. In the Global South, academics, activists and officials voice a radical criticism of the inhumane treatment of inmates suspected of terrorism in Guantanamo Bay as well as others in US supermax prisons; but at the same time, the same draconian tactics of torture designed and practiced by the US and the UK are adopted by their governments. Demirsu argues this happens because the fear of terrorism creates a climate of political instability in the Global South that may affect the credibility of officialdom. One of the chief goals of the nation-state aims at protecting citizens through deployment of repressive means which are legally legitimated. Since terrorism, which appeals to an illegal violence, hides in the population, some states use torture to gather vital information to prevent the next blow. The point of entry of this discussion seems to be that the advance of terrorism as well as the repressive counter-terrorism policies leads sooner or later to threaten basic rights and democracy. One of the most troubling aspects that precedes violence seems to be the state of exception, in which case, the government is given wide powers to cope with a temporary, major threat. If the state of exception is not regulated, the executive branch runs the risk of intervening in independent branches of government, the legislative and judicial, and thus harming the essence of the checks-and-balances powers.

Most likely there is no better example of how a State can threaten individual rights than Nazi Germany. The state of exception given to Nazis which resulted in the concept of Superman, as well as the decision of other European states not to intervene early in their internal politics, allowed the horrendous crimes against civilians and other ethnic minorities. Without the declaration of war, Demirsu adds, human rights or at best the well-being of citizens is monopolized by the State. At a closer look, this suggests that the discourse of human rights collides directly with the doctrine of the State's sovereignty. Differences aside, some voices claimed the risks of counter-terrorism policies to threaten the individual human rights of some minorities, while the process of securitization not only justifies such violations, but also introduces a much deeper logic of instrumentality where the end justifies the means. It seems that in modern Western democracies, the norms and rules lagged behind the need for further security. This is exactly what happened in the United Kingdom and Turkey, two notably different nations that share similar experiences regarding terrorism (Demirsu 2017).

Although there are many studies which proclaim on the intersection of human rights and terrorism, no less true is that Latin America (through the articulation of the Condor Plan) was gradually imbued into a climate of hostility and violence that led towards the suppression of all individual liberties (Dinges 2005; Jelin 2007; McSherry 2012). In this respect, as professor Skoll (2016) puts it, the red-scare, which derived from the Cold War, not only domesticated the wayward workforce within the US but was also exported to Latin America during the 1970s. Argentina, like Chile and Uruguay, was whipped by one of the cruelest and bloodiest dictatorship of its history, when the Government of Isabel Martinez de Peron was overthrown on March 24, 1976. A military Junta formed by General Jorge Rafael Videla, admiral Emilio Eduardo Massera and Brigadier-General Orlando Ramon Agosti took power starting with the "National Reorganization process". To some extent, though repression was orchestrated earlier than the coup, the process claimed thousands of disappeared dissidents as well as an array of serious human rights violations, a point which remains open to date (Kekes 1993; Slack 1996; Timmermann 2014; López 2016). In this section, we shall discuss in depth two books which take the problem of military repression as their primary object of study: *Genocide as a Social Practice* by Daniel Feierstein and *Game without End* by Jaime Malamud-Goti. Both positions sound pretty different though some complementary bridges can be constructed.

Let's start with Daniel Feierstein and his work. The Argentinian sociologist brings reflection on the dynamics of genocides against civilians narrowing the case of Nazi Germany to Argentina. Quite aside from the polemic argumentation, this book deserves some attention. Basically, as Feierstein acknowledges, the project initiated by the Process of National Reorganization (PNR) enthusiastically adopted the values of Western and Christian civilization, in which case, communism was its major threat. Following the contributions of Bauman, who envisaged genocides as the residual aftermaths of Western rationality, Feierstein (2017) alerts that beyond the degree of violence on the victims, genocides should be socially deciphered and studied according to a preceding ideological discourse. Through the articulation of genocides certain discourses and identities are imposed while others simply eradicated. Genocide should be understood firstly "as social practices" oriented to efface and reorganize the social scaffolding. The book explores not only the historical background of the genocide, which dates back to the end of WWII, but the main limitations of nation-states to protect their citizens. Paradoxically, a great number of slaughters are often perpetrated by the same nation-state. To the physical disappearance of peoples, which constitutes one of the most appalling crimes, Feierstein adheres, comparative studies of genocides suggest that some "discursive frames" precede the violence, emphasizing the subhuman condition of victims as the main line of justification of these horrendous acts. He starts from the premise that it is hard to narrow Nazis with PNR but insists,

> Historians who focus on the ethnic racism of Nazi Germany have tended to ignore the Nazi stereotype of the Jewish Bolshevik. The few historians that have tried to account for it have tended to subordinate the Nazis' ethnic racism to their ideological struggle against communism. However, the figure of Jewish Bolshevik was an unusual concept which merged the political and ethnocultural into a unitary image. The Jewish Bolshevik was portrayed as the prototypical enemy of Western civilization. The need to defend the West – or more exactly "Western Christian" values – would be invoked again much more explicitly as a justification for genocide in Argentina.
>
> (Feierstein 2017: 6)

However, Feierstein's argumentation rests on two main aspects. On one hand, genocides should be contemplated as an instrumental engineering disposed to impose a creative destruction, which means the reformulation and intervention of the ruling elite to reorganize the dynamics of other groups. On the other, these events, which are frequently state-led, need a technology of power to be successfully achieved. One point needs further attention: concentration camps in Nazi Germany were constructed to isolate first and exterminate later what they considered as undesired citizens. The same applied with PNR and the clandestine places of detention. In either case, the infra-valorization of the doomed "class", "group" or "ethnicity" as well as the process of demonization of the enemies of the nation was the key factor that led Juntas to perpetrate the genocide. Why does Feierstein use the term genocide in the case of Argentina?

As he contends, the term was coined by Lemkin combining the Greek *Genos* (tribe or race) with Latin *cide* (killing). What Lemkin looked for was a concept to describe something the world never witnessed before the rise of Hitler to power. The crimes against Jews not only were acts of cowardice because they were non-combatants but an endeavor to exterminate an entire ethnic group without any valid reasoning. This seems to be the first obstacle Feierstein faces. Since Juntas never imagined a racial project, why may their crimes be compared – in Feierstein's accounts – to those of Nazis?

Feierstein goes on to answer that the definition of genocide is based on any deliberate attempt to kill ethnical, racial or religious groups according to their thinking or any constitutive feature. For the moment, the legal definitions anyway are broader and sometimes escaping the principle of equality before the law. The Juntas disposed and devoted the resources of the State to pursue and kill dissidents, who were assassinated without a fair trial. Most certainly, as he writes,

> Bearing all this in mind, I will define a genocidal social practice as a technology of power—a way of managing people as a group—that aims (1) to destroy social relationships based on autonomy and cooperation by annihilating a significant part of the population (significant in terms of either numbers or practices), and (2) to use the terror of annihilation to establish new models of identity and social relationships among the survivors. Unlike what happens in war, the disappearance of the victims forces the survivors to deny their own identity—an identity created out of a synthesis of being and doing—while a way of life that once defined a specific form of identity is suppressed. Accordingly, I will use the term "genocidal social practices" to distinguish these specific processes from the legal concept of genocide.
>
> (Feierstein 2017: 14)

With these connotations in mind, Feierstein says that any practice oriented to destroy systematically a group's identity should be deemed as a genocide practice, simply because it was the original meaning Lemkin endorsed to the term. Equally important, Juntas not only orchestrated a repressive apparatus beyond the law, but also incurred extreme violence to silence some critical voices. Though we agree with Feierstein this cannot be glossed over, some points need further clarification.

Firstly and most important, Feierstein does not explore terrorism and human right violations as two sides of the same coin, nor the role of democracy in the formation of radicalized cells such as Tupamaros, ERP or even Montoneros. The political instability that cemented the "dirty wars" (a term used by some scholars such as Suarez–Orozco, 1987; Pinet, 1997; Portela, 2009; and Kohut & Vilella, 2016, among many others) was not determined by the military coup, but it was encapsulated in the core of democracy. As James Piazza brilliantly observed, one might speculate that consolidated democracies are immune to terrorism but it is a clear mistake. The UK faced Irish guerrillas in the same way that other democracies did. The fact is that terrorism emerges when one of the parties or factions pass to clandestinity, abandoning representation in parliament.

This leads to thinking that terrorism and human rights violations can be enrooted in democracy (Piazza 2007, 2008). Piazza's contributions interrogate critically to what extent democracy immunizes us to the scourge of terrorism. Secondly, as Kekes showed, the motivations of Juntas and Nazis were pretty different while both processes have some commonalities. While the Nazis imagined a racial empire based on the total obliteration what they dubbed as "sub-humans", in which case, this could be performed through impersonal and instrumental methods, army forces have strongly believed they were doing the correct thing struggling against communists. The difference lies in the emotionality that the army forces in Argentina evinced, Kekes (1993) concludes. Both committed horrendous crimes but going along different paths. Such remarks open the discussion into other twilight horizons which were not addressed by Feierstein in his text. Lastly, while we think of genocide as an attack against defenseless groups, by the stronger agent we are forced to assume, the feature of the groups plays a vital role in configuring the borders and the essence of such a crime. This means army forces are guilty of genocide because of their constitutive traits as a privilege group with respect to victims. At a first glimpse, this assertion equals a breach with the nature of Roman law, where offenders are only judged by their crimes not by their status. Of course, this is not easier since the Juntas monopolized the power of the State to vulnerate – whether guilty of crimes or not – lay-citizens.

The second author to review is Jaime Malamud-Goti who in 1996 published *Game without End: State Terror and the Politics of Justice* (Oklahoma University Press). In sharp contrast with Feierstein and retributism, which is an academic wave convinced that criminals should be punished regardless of the effects of the punishment. In the case of state terror, as he observes, analysts should make closer scrutiny of the fact that putting military forces on trial poses serious risks for democracy. The retributists, as Malamud-Goti cites, ignore the contextual nature of crimes as well as the constraints over where retributism operates,

> Retributists who have advocated the punishment of state criminals have generally disassociated such a punishment from its consequences. The message to the wrongdoer is, this is how wrong your crime was. Retributive factors – the harm done or the culpability of the actor – are seen as constraints on the treatment of individuals, and as tools for the promotion of social and the state's interests… the value of retribution is simply that, by disregarding the effects of punishment, it does not indeed place constraints on society. For a full blooded retributist, punishment is demanded of every military officer who participated in violations of human rights, even if the consequences are a military revolt.
>
> (Malamud-Goti 1996: 12)

This cited excerpt does not mean, the author clarifies, that the Juntas should not be punished as criminals, but far from being extended to all army forces members, the trial should be executed through the heads or the spin-doctors who carefully planned the acts. Otherwise the aftermaths of the trials may very well be count-er-productive and very harmful for democratic life.

Malamud-Goti introduces a more than interesting division between "full-blooded retributists" who seek to punish all involved officials without evaluat-ing the consequences, and "goal oriented retributists". While the former appeals to the doctrine that "all guilty perpetrators should be punished" (Malamud-Goti 1996: 13), the latter embraces a contextual explanation of facts that implement a selective punishment applicable to the top-ranked officials. This starts from the premise, to which Malamud-Goti adheres, that at the time some activist groups – like Madres de Plaza de Mayor – adopted a utilitarian perspective on the punish-ment to military forces, it places the democracy in jeopardy. Raúl Alfonsin faced four attempted coups that revealed the fragility of Argentinian political institu-tions. Each involved group developed its own conception of the extreme violence Argentina went through in the 1970s, constructing its own interpretation of his-tory. This does not mean to say overtly that military officers should be blamed or put on trial, but the obsession for introducing retributism paved the way for some reactionary discourses oriented to justify the Juntas' repressive action.

> The thesis of this book does not lie on the assumption that military officers lacked moral and legal responsibility, rather, it is the opposite. It claims that responsibility was more spread out than the populace is ready to acknowl-edge. In maintaining that the human rights trials served a generalized reluc-tance to accept responsibility, I imply that formalized blame contributed to the unchanged situation of Argentina vis-à-vis attitudes to terrorist power.
> (Malamud-Goti 1996: 25)

In this stage, Malamud-Goti acknowledges that though the intentions of civic judges were in consonance with the democratic ideals, the effects went in the opposite direction. On one hand, military officers were under oath to accept civil-ian power. On the other, the society was divided in two: those who claimed that the terror-state perpetrated acts of genocide against civilians which need to be expiated and those who thought there was a dirty-war and the violence came from the two bands. The retributists, an ideological movement which is based on utili-tarianism – judging an act by its effects and not its reasons – emphasizes the needs of putting all involved offenders on trial in order to legitimate the civic order. The legitimacy of law can be honored only when crimes are punished. However, this is not supported by other voices that alert that groups tend to blame others for their own crimes and in Argentina the real contexts, where the Juntas and dirty-war operated, are far from being unveiled. It is particularly interesting that the involved factions remain with a partial viewpoint of the state-terrorism which obscures more than it clarifies. Malamud-Goti coincides with Feierstein: many junior officers were educated and trained to think they were a key instrument in the struggle against the infiltration of communism in the country and the *guer-rilleros* (terrorists) were sub-humans. In any case, responsibility and punishment are not only social constructs which should be negotiated among involved parties but also vary with time and contexts (Malamud-Goti 1996). The Juntas not only committed crimes appealing to much deeper nationalism and the demonization of dissidents, they politically manipulated the language to pose a specific meaning of "subversion". Supported by the Catholic Church, the Juntas developed a closed discourse where all dissidents or external nations which can deliver some critical speech against the human rights violations in Argentina, were labeled as "enemies of the nation". In that way, military forces escaped any moral charge because from their viewpoint they were struggling against those who threatened the order of the Republic. Meanwhile, lay-citizens and rank-and-file workers were not only seduced by Videla's rhetoric but also succumbed to the action of terror which played a paralyzing role pushing them back from the public sphere. People were terrorized and the effects were isolating and paralyzing (Malamud-Goti 1996).

Malamud-Goti holds the thesis that societies go around three different conceptions of power, which are orchestrated according to each context: articulating, disarticulating and structural power. The articulating subtype refers to life in a democracy where social instructions coordinate activities under the auspices of the law. However, under some conditions these rules are suspended and the disarticulating rule emerges. The Juntas governed through a disarticulating power which aimed to prevent social coordination. As a result, people were paralyzed and their frustration and fear were re-channeled towards a political immobility. The state of terror made panic a political style, paving the way for a "structural power" which transcends even when the regime is ousted. Hence, Argentinians went through pathological forms of understanding the "Other", which remained even once democracy returned. In the mid-1970s, the "red-scare" was the justification while terror was the instrument of domestication that radically changed the ways Latin Americans imagine politics. Whatever the case may be, military forces invoked the rights of liberality to annihilate what they considered an irreversible evil (communism) in the same way that contemporary society remains indifferent to the human rights violations in Supermax prisons. Capitalism seems to be inextricably intertwined with the right of instrumentality, which poses the ideals of *the means justify the ends*. This point leads us to reconsider the nature of democracy and its legal resources to prevent the rise of dictatorship. While potentially Feierstein echoed his great sensibility as Marxist and Jew (he uses these terms in his text), two of the preferred targets of right-wing nationalism that empowered the Juntas' decision-making process, Malamud-Goti served as advisor (in human rights fields) for Raúl Alfonsin's presidency. From different angles, both positions illustrate the worries of the different arguments over the crimes perpetrated by the Juntas, but not only are Argentinians reminded of these horrendous times but also the ethical dichotomies revolving around Western democracies.

Rethinking democracy

Most likely, Cornelius Castoriadis (2006) would be a marginal voice in the constitutional theory of American studies in English speaking countries, but undoubtedly he had laid the foundations of an all-encompassing model to understand the role of ancient Greek mythology in the formation of democracy. He writes that Greek tradition cannot be understood without exploring Homer's legacy. Greeks not only cultivated astronomy and mathematics, but also practiced a politics that was not shared by any other culture of their time, democracy. For Hellenes, the spirit of Greece was democracy and the right of legislation. The essence of mankind is reflected in the possibility to sanction and abide by the law. In the Odyssey Homer situates Ulysses as stranded in the land of the Cyclops. Once there, Ulysses describes their customs and daily habits as appalling, proper of monsters without law or the organization of assemblies. Under the figure of monstrosity surely Homer narrated the habits of undemocratic cultures which were unfamiliar with the possibility of legislating their own fates. It is important not to lose sight of the process of decision making as the touchstone of Hellenics and their civilization. However, in any case, democracy was neither a consolidated term nor a form of government. As Castoriadis explains, for ancient Greeks the concept of *moira* means the immanency of death for all beings. Even the gods (in their immortality) were not beyond the action of *moira* (fate). Destiny encompasses everything in the Homeric tradition but mysteriously not the law. One of the characteristics that separates Greece from the rest of ancient mythical structures is the lack of revelation and prophecies about the future. Since Greek mythology does not refer to a world created for humans, they comprehend that the body of law is the only instrument capable of giving order in politics. Even though predestination and divination were two widespread customs in order for solicitants to conduct business or face certain threats, nobody in Greece would have consulted these techniques to promulgate the laws. From this perspective, Castoriadis dwells on those points that outline the main heritage of Ancient Greece. Among the contributions of this civilization we find the agonal competition for glory and fame, the quest for trust, the tension between essence and presence (*doxa* and *nomos*) and finally a determination for democracy. Here a point that merits a certain degree of consideration: what is the relation between fate and competition?

In ancient cultures the king not only kept absolute power, but his legitimacy was hereditary as in modern monarchies. The only difference with other monarchs was the use of *Demos*, a legal resource where the king asked the assembly for advice (when he was unsure) or simply a lay-citizen wanted to derogate some laws. This is the root, Castoriadis reminds us, of democracy, which has nothing to do with modern democracy.

Last, Robert Castel examines painstakingly the alterations suffered by the social bond from medieval times to modernity. The political freedom was conducive to the process of industrialization in late medieval times. The sense of liberty not only facilitated internal migration directly to cities, but also situated farmers in a new, unknown condition, the possibility of choice. As a result of this, many poor peasants were free to negotiate with more than one landowner about their conditions of subsistence. The capital found the necessary mechanism, articulated by the doctrine of the liberal market and democracy to sustain the basis of the nation-state. Whether the democracy in old Greece opened the possibility of any citizen to derogate a law if necessary, Anglo-democracy focuses on the voting process but creates a gap between citizens and their institutions. This hole is filled by the financial powers that operate inside the nation-states. That way, the individual voices in democratic regimes were trivialized in view of the hegemony of jurisprudence and legal theory. The laws sanctioned in modern parliamentary democracies seem not to be objected to by citizens. The individual views fail against the corporate powers that monopolize the well-functioning of the republic. Therefore, it is contradictory to attack the market's defense of democracy. The capitalist logic today, enrooted in modern Anglo-democracy, is not being placed under the lens of scrutiny. The functioning of Anglo-societies today depends on their efficiency to promote conflict in the periphery. The concept of Anglo-democracy also represents the idea of republicanism, given not as a form of political commitment, but as a way of forging control over the institution of society. The order for the UK and US is based on the right of property as well as the degree of control that the elite have to dominate citizenry. The Anglo-democracy denotes a new type of organization, unknown until the British Empire flourished. Although the electorate votes for their authorities every four years, a question that characterizes the democratic spirit of Law, they are impeded from derogating the laws sanctioned by the senate. This leads modern society to corporative forms of democracy where the will of individuals are systematically violated. The liberal myopia in denouncing this depends on the prejudice that the safe republic is formed only when the rights of people are granted. These rights are based not only on the freedom disposed for hyper consumption but also in the obsession for property ownership. Anglo-democracy is after all, a deformation consolidated by the end of WWII.

Conclusion

One of the merits of British Empire and the US in forging liberalism as an ideological platform of control alludes to the efficacy in controlling a vast empire which had no resistance to imperial rule. While London disposed free trade, which supposedly led towards a virtuous world, little was said on the cruelty and the state of exploitation of British Army forces in the peripheral colonies. This was the reason why (Anglo) democracy flourished in these densely populated cities. Marking the homeland as an exemplary center, the ideology legitimized the action of Empire abroad. The same applies to the United States, a young nation that continued the tactics of politics and subordination of London. Since underdeveloped nations are unable to fix the rules of trade globally, they appeal to arbitrary policies that sometimes lead to dictatorship or autocracy. This chapter showed amply not only how the sense of democracy distanced itself from the Hellenic spirit, but how even in democracy the framing of rules can be altered according to the preservation of the status quo. The ideological core of the liberal, state-educated workforce hides the invisible hand of exploitation while the need to belong to the club of chosen peoples is cultivated. Lay-citizens rarely question the actions of their government in other countries, because they feel they are good people paying taxes, sending children to school and voting. However, unlike ancient democracy where rights and laws can be derogated by the assembly, in Anglo-democracy there is a gap between citizens and the social institutions, which is filled by the political corporation. In ancient Greece, elections were never an option though people had access to revert those laws that were deemed as unjust. It is important to add that these types of government are applicable to small populations; Anglo-democracy and the introduction of the liberal state homogenized scattered populations under the limits of borderlands.

References

Almond, G., & Verba, S. (1992). Civil culture and stability of democracy. *Polis. Political Studies*, *4*(4): 122–129.

Altheide, D. (2017). *Terrorism and the Politics of Fear*. New York, Rowman & Littlefield.

Anbinder, T. (1992). *Nativism and Slavery: The Northern Know Nothings & the Politics of the 1850s*. Oxford, Oxford University Press.

Bailyn, B. (2017). *The Ideological Origins of the American Revolution*. Cambridge, Harvard University Press.

Baker, L. D. (1998). *From Savage to Negro: Anthropology and the Construction of Race, 1896–1954*. Berkeley, University of California Press.

Bay, C. (1992). "Politics and Pseudo-politics: A Critical Evaluation of Some Behavioural Literature". In *Approaches to the Study of Politics*, B. Susser (ed.). New York, Macmillan Co., 51–75.

Castoriadis, C. (2006). What Shapes Greece: From Homer to Heraclitus. Seminaires 1982-1983. Human Creation II. Buenos Aires, Fondo de Cultura Economica.

Dahl, R. A. (1992). "The Behavioral Approach in Political Science: Epitaph for a Monument to a Successful Protest". In *Approaches to the Study of Politics*, B. Susser (ed.). New York, Macmillan Co., 27–48.

Demirsu, I. (2017). *Counter-Terrorism and the Prospects of Human Rights: Securitizing Difference and Dissent*. New York, Palgrave Macmillan.

Dinges, J. (2005). *The Condor Years: How Pinochet and His Allies Brought Terrorism to Three Continents*. New York, The New Press.

Donohue, K. G. (2003). *Freedom from Want: American Liberalism and the Idea of the Consumer*. Baltimore, Johns Hopkins University Press.

Durkheim, E. (2014). *The Division of Labor in Society*. New York, Simon and Schuster.

Easton, D. (1992). "Tenets of Post-behaviouralism". In *Approaches to the Study of Politics*, B. Susser (ed.). New York, Macmillan Co., 49–50.

Feierstein, D. (2017). *Genocide as Social Practice: Reorganizing Society under the Nazis and Argentina's Military Juntas*. New Brunswick, Rutgers University Press.

Guidotti-Hernández, N. M. (2011). *Unspeakable Violence: Remapping US and Mexican National Imaginaries*. Durham, NC, Duke University Press, 374.

Haggard, S., & Kaufman, R. (1995). *The Political Economy of Democratic Transitions*. New Jersey, Princeton University Press.

Harris, M. (2001). *The Rise of Anthropological Theory: A History of Theories of Culture*. Walnut Creek, AltaMira Press.

Heilbroner, R. L. (2011). *The Worldly Philosophers: The Lives, Times and Ideas of the Great Economic Thinkers*. New York, Simon and Schuster.

James, T. S. (2012). *Elite Statecraft and Election Administration: Bending the Rules of the Game?* Basingstoke, Palgrave Macmillan.

Jelin, E. (2007). Public memorialization in perspective: Truth, justice and memory of past repression in the southern cone of South America. *The International Journal of Transitional Justice*, *1*(1), 138–156.

Kekes, J. (1993). *Facing Evil*. Princeton, Princeton University Press.

Kohut, D., & Vilella, O. (2016). *Historical Dictionary of the Dirty Wars*. New York, Rowman & Littlefield.

Korstanje, M. E. (2017). *Terrorism, Tourism and the End of Hospitality*. New York, Springer Nature.

Lewis, D. (1973). Anthropology and colonialism. *Current Anthropology*, *14*(5), 581–602.

Limerick, P. N. (1987). *The Legacy of Conquest: The Unbroken Past of the American West*. New York, WW Norton & Company.

López, F. (2016). *The Feathers of Condor: Transnational State Terrorism, Exiles and Civilian Anticommunism in South America*. Newcastle upon Tyne, Cambridge Scholars Publishing.

Malamud-Goti, J. (1996). *Game without End: State Terror and the Politics of Justice*. Norman, Oklahoma University Press.

Martin, J. G. (1964). Racial ethnocentrism and judgment of beauty. *The Journal of Social Psychology*, *63*(1), 59–63.

McSherry, J. P. (2012). *Predatory States: Operation Condor and Covert War in Latin America*. New York, Rowman & Littlefield.

Murillo, S. (2008). *Colonizar el Dolor*. Buenos Aires, CLACSO.

Olson, M. (1971). *The Logic of Collective Action: Public Goods and the Theory of Groups*. Cambridge, Harvard University Press.

Pels, P. (2008). What has anthropology learned from the anthropology of colonialism? *Social Anthropology*, *16*(3), 280–299.

Piazza, J. A. (2007). Draining the swamp: Democracy promotion, state failure, and terrorism in 19 Middle Eastern countries. *Studies in Conflict & Terrorism*, *30*(6), 521–539.

Piazza, J. A. (2008). Do democracy and free markets protect us from terrorism? *International Politics*, *45*(1), 72–91.

Pinet, C. (1997). Retrieving the disappeared text: Women, chaos & change in Argentina & Chile after the dirty wars. *Hispanic Journal*, *18*(1), 89–108.

Portela, M. E. (2009). *Displaced Memories: The Poetics of Trauma in Argentine Women's Writing*. Lewisburg, Bucknell University Press.

Raventós, C. (2008). *Democratic Innovation in the South. Participation and Representation in Asia, Africa & Latin America*. Buenos Aires, CLACSO.

Rawls, J. (1999). *The Law of Peoples*. Cambridge, Harvard University Press.

Said, E. (1979). *Orientalism: Western Representations of Orient*. New York, Vintage.

Shue, H. (1996). *Basic Rights: Subsistence, Affluence and US Foreign Policy*. Princeton, Princeton University Press.

Simon, J. (2007). *Governing through Crime: How the War on Crime Transformed American Democracy and Created a Culture of Fear*. Oxford, Oxford University Press.

Skoll, G. R. (2016). *Globalization of American Fear Culture: The Empire in the Twenty-First Century*. New York, Springer.

Skoll, G. R., & Korstanje, M. E. (2013). Constructing an American fear culture from red scares to terrorism. *International Journal of Human Rights and Constitutional Studies*, *1*(4), 341–364.

Slack, K. M. (1996). Operation Condor and human rights: A report from Paraguay's archive of terror. *Human Rights Quarterly*, *18*(2), 492–506.

Suarez-Orozco, M. M. (1987). "The Treatment of Children in the 'Dirty War': Ideology, State Terrorism and the Abuse of Children in Argentina". In *Child Survival*. Dordrecht, Springer, 227–246.

Sugden. R (2005). *The Economics of Rights, Cooperation and Welfare*. London, Palgrave Macmillan.

Susser, B. (1992). *Approaches to the Study of Politics*. New York, Macmillan Co.

Timmermann, F. (2014). *El gran terror*. Miedo, emoción y discurso. Chile, 1973–1980. Santiago de Chile, Copygraph.

Weber, M. (2002). *The Protestant Ethic and the "Spirit" of Capitalism and Other Writings*. New York, Penguin.

2 Neoliberalism, consumption and poverty

Introduction

In this chapter, I confront two contrasting theories: liberalism and Marxism. This task – needless to say – was daunting and took much time and effort. Of course, because of space, it is almost impossible to review all published books and studies on the economy of liberality or radical materialism but, more importantly, we place in dialogue two senior lecturers who have dedicated their lives to the study of capitalism (Zygmunt Bauman and Kathleen Donohue).

It is no accident that liberalism assumes that self-interest regulates the social life as the best judge of what is pertinent for citizens. The variants and liberties gained by the citizens reach their zenith in the exchange of the market. For liberals, the State should intervene only when an extreme conflict and irreconcilable positions arise (Dryzek, Honig & Phillips 2008). Hence the liberal state – as validated in the introductory chapter – combines two important values. On one hand, it orchestrates an internal sense of freedom which is directed to prosperity, consumption and wealth production. If the ideology of Enlightenment proved something, it related to the use of rationality as the only truth to pursue. This point marked a radical division between Europe and the rest of the world. The Western rationality played a vital role in the configuration of contemporary society and modern science. However, at a closer look, one might speculate that this sense of objectivity not only is socially constructed but undermines the real foundations of democracy (as denounced by Feyerabend). The imposition of the truth not only can be counterproductive for open democracy but it can also be politically manipulated by the ruling elite to preserve their status quo. Equally important, as a project originally created to be expanded, the Enlightenment – voluntarily or not – conditioned the development of a conception of evolution, which gradually crystallized into "social Darwinism". Adjoined to puritanism, social Darwinism fostered a climate of competition and solipsism where workers – instead of cooperating with others – devoted their efforts in showing they were part of the chosen peoples. This mythical discourse not only molded the contours of economic theory but also legitimated the adoption of consumption as the milestone of the capitalist economy. The society of producers which characterized the mercantilist ethos and the industrial world set the pace for the advance of a society where consumption was the priority. The quest for diversity and the adoption of cultural consumption were two important tenets of neoliberalism, as this chapter opportunistically shows. The thesis here is that the culture of Thana-Capitalism depends upon a previous stage, which is given by social Darwinism and the premises of the Enlightenment. To a major or minor degree, different voices have pointed out that capitalism has more lives than a cat, alternating new facets once a threat is faced. Quite aside from this, the society of consumers started from a fear of poverty, but poverty was never reverted. Capitalism made a spectacle of the external risk and in doing so, it is internalized. Naomi Klein wrote that capitalism recycles from its own catastrophes, replicating not only the material asymmetries that facilitated the disaster but also cementing the power of elite, which is never questioned (Klein 2007). To some extent, the system never corrects the glitches but only commoditizes them in the form of spectacles. Of course, Thana-Capitalism offers "a spectacle of disaster" which, far from correcting the unjust economic backdrop, legalizes it (Korstanje 2016). As Donohue outlined, the fear of poverty cemented the passage from a producerist to a consumerist society, but poverty remains. Likewise, the society of consumers sets the pace for a new stage of capitalism, Thana-Capitalism, where the spectacle of disasters and the obsession for gazing at the Others' death prevail. Eternal life is still impossible but through the Others' death, modern citizens maximize their (sadistic) pleasure. This is a particularly big problem while we glimpse the advance of modern terrorism and its interplay with media.

Formal standards of liberalism

In order to gain further understanding on liberalism, we have to delve into the liberal mind, a point which was widely studied by Professor Emeritus Richard H. Pells. In his book, *The Liberal Mind in a Conservative Age*, he confirms that though originally liberals were enthusiastically embracing the concept of liberty as their mainstream value, the borders of liberalism, as well as their ideological core, varied from decade to decade. From the beginning of WWII, academicians from liberal stripes agreed to support the Soviet Union in the struggle against the tyranny of Adolf Hitler. However, once the beast was decapitated, several thinkers such as Arendt, Schlesinger and Burnham endorsed their support for Harry Truman instead of Stalin. In the decades to come, Marxism in a broad sense and especially communism were catalogued as the main threats to liberal Western democracies. For some reason, liberals not only adopted the cult of political personality but also faced serious problems in achieving an academic autonomy respecting officialdom. Whatever the case may be, liberalism and democracy were two key factors that accompanied American sociology from the outset, even now. The rise of McCarthyism, while criticized by the most influential, authoritative voices, reinforced very well the anti-communist sentiment, which was historically cultivated in the American hearth. Joseph McCarthy inaugurated a new conservative age leading liberal writers to theoretical Marxism, which was helpful to understand the problems and state of exploitation that the workforce suffered (Pells 1989). To some extent, as E. Hobsbawm (1995) puts it, the history of the twentieth century witnessed a radical politics that veered from one extreme to the other. This suggests that communism and liberal capitalism were intractably intertwined. While liberal governments ruled the destiny of the United States, academicians found in Marx and Marxism a guideline to follow, even in times of stock and market crises; but once the Cold War emerged, democracy was weighted as the best of the feasible forms of governments. As a result of this, the fall of the Soviet Union and its promises accelerated serious problems for capitalism laying the foundations for a new left-wing intelligentsia which operated within the critical lines of sociology and social sciences. This begs an intriguing question: to what extent are democracy and liberty the ideological side of an oppressive capitalism?

To address this question, it is noteworthy that the configuration of nation-state contains the essential background which historically articulated liberal thinking with a culture of mass-consumerism. The book *Liberal Loyalty*, which is authored by A. Stilz (2009), seems to be of paramount importance to elucidate some answers to the doubt casted above. She coins the term "liberal reasoning" to describe the loyalties of citizenries to their states. From Hobbes to Levi-Strauss, thinkers have theorized on the factor which keeps society united. While it is often possible that some discontents will emerge, framers turned their attention to developing the necessary conditions to scrutinize citizens while at the same time the conditions of frustrations were duly regulated if not reversed. Stilz interrogates critically the extent to which we can blame US or UK citizens for the military actions performed by these two nations in the Middle East. Are elections a valid form of renovation or simply an ideological justification to blame others for our passivity?

In the midst of this mayhem, Stilz acknowledges that the citizens of modern democracies are simply educated to think that while they work and pay taxes, they are "good boys". This logic leads to a philosophical dilemma because citizens are unable to prevent the passing of unjust laws. The Hobbesian conception of State signals the doctrine of security as the platform to which all citizens agree; collective rights are redeemed in view of a much broader goal. Stilz proffers an interesting model to understand the loyalties of modern citizens to nation-states, escaping the ethical burdens of what politicians do. For the sake of clarity, nationality plays a leading role limiting the loyalties of individuals to specific law-making. Lay-citizens should abide by their laws but only when they stay in their native soil. While traveling, they are subject to new jurisdictions and laws. As Stilz confirmed, the concrete discourse of liberalism succumbs when we imagine situations where some unjust laws, which are passed in parliament with a majority, should be abided. The principle of redistributive justice not only does not work, but also becomes counterproductive. If lay-citizens are morally pressed to obey a new emerging dictator (such as Hitler or Stalin), how do they supposedly behave? Are they liable for the political crimes of their new regime or simply companions of such immoral acts?

The ideology of enlightenment

Before starting this section, I wonder whether it is possible to know more about the intersection of Enlightenment and the sense of reality with the world of consumption in contemporary society. At a first glimpse, the world seems to be something else that we (humans) often imagine. As Tim Ingold puts it, the idea of science can be seen as an invention of Western civilization and the notion seems to be unfamiliar to hunters and gatherers and other civilizations. In philosophical terms, this suggests that the word reality denotes some ethnocentric construction. Hence hunters and gatherers perceive their environment in a relational perspective which is connected to nature. We, the Westerners, feel the world in two radical contrasts: humans and the rest of creation. Over centuries, we have developed and improved a "dwelling perspective" which resulted in a specific cosmology. In this respect, science helps us not only understand the world but also introduced the (sacred) principle of rationality as a vehicle to produce and consume (even wasting) what we need to live but at the same time creating a surplus which later will be stored and exchanged with others (Ingold 2000). These are exactly the roots of the capitalist system. So science and capitalism seem to be inextricably intertwined. In his work *Dialectics in Social Thought,* Professor Geoffrey Skoll (2014) interrogates the nature of Western rationality as an ideological construction aimed at legitimating the power of the ruling elite. He holds polemically that, though never questioned, the sense of reality works in the same way dialectics do. The success of capitalism to be imposed as the only economic system consisted in manipulating dialectics to prevent social change. To put the same in other terms, as Skoll maintains, not only psychoanalysis but also Marxism needed a third object to validate their main doctrines. While the former appealed to the figure of the unconscious, the latter signaled to commodity fetishism. These constructions explained everything but at the same time explained nothing. Over years, social science constructed a third object (dialectics) in order for bourgeois society to domesticate the discontent of rank-and-file workers. As a third object, dialectics, has received considerable attention from experts and epistemologists, while the social order is given as a taken-for-granted certainty.

Richard Bernstein elicits an interesting review on the developments of different epistemologists respecting the nature of Enlightenment. Far from being open or democratic, these voices agree that there is an identity of rationality which becomes gradually oppressive against the individual. Here, as he notes, the paradox lies in the fact that the rationalization process opens the doors to its own destruction. An action can be defined as rational when it is oriented to purposive ends, which are carefully checked.

> We can now understand that if Weber is Right, if this is what rationality and rationalization have become in the modern world, if this is the inevitable consequences of the very type of emancipation which the Enlightenment fostered and legitimized, then we can well understand why there is a rage against reason.
>
> (Bernstein 1988: 200)

While the imposition of reason was a Western invention, no less true is that it was articulated to dominate wildlife, and in so doing, the man became the wolf of the man. The instrumentality of emotions and life not only leads humans towards an atmosphere of alienation but also constitutes the legal platform towards Auschwitz. In effect, it was not the banality of evil (paraphrasing Arendt) but the instrumentalization of the Other which facilitated the atrocities perpetrated by Nazi Germany. The notion of truth as fostered in the Occident is not only an ideological mechanism but a major threat for democracies (Bernstein 1988). As an epistemological project, liberalism falls into the contradiction of thinking its individualism is a positive value. Modern liberalism enthusiastically embraced as a governing theory that all individual liberties should be equally respected. However, at the time it is applied in social contexts, it fosters a competitive individualism that tries to be imposed as a universalist project (Schlosberg 2008).

A similar point of entry in this discussion was presented by Paul Feyerabend in his book *Science in a Free Society.* The separation of powers and the derived checks and balances not only cemented the republican tradition but also the principle of modern science. In democratic societies, science remains separated from the State and it divides the Church and religiosity. The germen of rationality is replicated through a complex system of education which cultivates "the scientific method" over other voices and traditions. Following this, Feyerabend contends, while science accepts the triumph of rationalism over religion, witchcraft or even astrology, it was finally trapped into a conceptual gridlock. Unlike experts, lay-people feel that science is the correct option simply because it employs the right method, but one thing the history of science shows seems to be that the rules that make rationality may be changed, negotiated or breached according to the interests of the ruling elite. Feyerabend exerts a radical criticism not only of rational thinking but also of the epistemology of science. From his viewpoint, science is contradictory to the real democracy, or at best the real cultural values of a free society (Feyerabend 1982). Hence this suggests that contemporary society "should be defended" from science. He starts a seminal book chapter, which is included in *Philosophy: Basic Readings* (a book edited by Nigel Warburton), in the following manner,

I want to defend society and its inhabitants from all ideologies, science included. All ideologies must be seen in perspective. One must not take them too seriously. One must read them like fairytales which have lots of interesting things to say but which also contain wicked lies, or like ethical prescriptions which may be useful rules of thumb but which are deadly when followed to the letter.

(Feyerabend 2012: 261)

As a cultural project, science was thought – in the days of Enlightenment – to be a necessary instrument against "superstition" and authoritarianism, a radical reaction of the liberal mind, Feyerabend adds, to the arbitrariness of religiosity. Basically, many writers and philosophers have adamantly criticized the life of contemporary community but leave science out of the discussion. Science, after all, would be the best antidote to the irrationality of tyranny. Through the articulation of different discourses, the sense of truth imposes on the dark veil of religion, which looks to captivate and dominate the free peoples. Nothing of this was true, he scornfully writes. If democracy was conceived to be contrasted against authoritarianism, science played a leading role in further illuminating the laws of nature. From its outset, modern science demythologized the world (Feyerabend 2012). Most certainly, although truth would be sounded as a neutral word, it is an ideological instrument of control. As he comments, modern reasoning, in some way, prevents the real liberty of citizens to think freely. In fact, if I think in a factual basis, the truth appears to be the negation of individual choice.

> My criticism of modern science is that it inhibits freedom of thoughts. If the reason is that it has found the truth and now follows it then I would say that there are better things than first finding, and then following such a monster.
>
> (Feyerabend 1982: 263)

The above noted concern is one of the contributions of Feyerabend to our debate and of course his legacy to the epistemology of science. It is important not to lose sight of the sense of instrumentality gradually crystallized in the project initialized in the nineteenth century, which paved the way for the rise of Thana-Capitalism (a theme to be developed in the next chapters), *social Darwinism.*

With the benefit of hindsight, social Darwinism evinces the triumph of rationalism over other forms of thinking, in which case the survival of the fittest is situated as the main cultural value of capitalism. Selection and not free choice is the main right that sorts the human world in the same way that it controls the wildlife. In consequence, we, the citizens, should struggle to demonstrate that not only are we special but we are in the right to live (Hofstadter 1944).

French philosopher Paul Virilio departs from this point in his book *The University of Disaster.* He argues convincingly that far from being oriented to help others, Modern Science was defeated by the interests of bourgeois society and capital-owners. He continues the anarchical view of Feyerabend, but from a different angle. The introduction of technology was originally conducive with the needs of making this world a safer place. The concept of well-being was the touchstone of modern science in former centuries. However, these values were radically altered for a profit-centered logic which escapes any humanism. The exegetes of science are not interested in protecting the more vulnerable citizens or improving the conditions of life, they are financially supported to perpetuate the basis of exploitation which marks a gap between elite and workforce. To validate his thesis, he uses the example of global warming. Scientists working 24/7 at their laboratories are not moved by reversing the negative effects of global warming but only in mitigating the negative effects so that capitalism can survive. The world is like a great air-conditioning system in a closed environment. We are not concerned by the rise of temperature; we close in on our sacred bubble surround of high tech. This happens because scientists abandon critical thinking in view of further financial investments (Virilio 2010). In that way, disaster results as a consequence of the depersonalization that technology accelerates. In this modern society, the needs of hyper-consumerism have not only replaced humanism but place genuine democracy in jeopardy, as was vaticinated by Paul Feyerabend.

Richard Hofstadter (1944) traces the origin of social Darwinism, an academic trend which postulated the rise of whites over other racial groups as well as the needs of imagining society as the sum of agents, all struggling with others to dominate. This issue will be addressed in the next section in some detail.

The conception of rationality as the pursuit of goals and ends was one of the milestones of modern democracies. According to the literature, people chose their candidate according to their preferences which express the sum of future gains and the avoidance of costs (Weingast 1979; Ostrom 1998; Blais 2000; Amadae 2003). In the seminal book *An Economy Theory of Democracy*, Anthony Downs reminds us of the dichotomy between the rational behavior of voters and the influence of their in-group and peers. Downs reviews different case studies to show that democracy shares a rational architecture which can be logically planned. The distribution of income has a direct impact on individual behavior as well as in-group interests. He overtly writes,

> Decision-making is a process which consumes time and other scarce resources; hence economy must be practiced in determining how many resources shall be employed in it. This fact forces decision-makers to select only part of the total available information for use in making choices. The principles of selection they use depend upon the end for which the information is a means.
>
> (Downs 1957: 218–219)

From the above excerpt one might speculate that within the logic of the liberal state, decisions are made on the basis of scarce information, which suggests that the sense of liberty is limited to the goods that circulate in the economic system. Besides, neither democracy nor elections work without the principle of selection, which is the touchstone of contemporary society (Downs 1957). One of the aspects that Downs did not mention is that rationality, in this vein, would not resolve the inequalities of the system, unless by the introduction of ignorance, which helped the ruling elite's policies to be legitimated. As Feyerabend explained, leaving the political organization to the rules of rationality implies not only coercing individual liberty because the subject is determined by the economic environment, but also imposing individual ends to affect the collective well-being. Nothing of this can be done without the introduction of selection as the main ideological discourse of capitalism. This begs the question: to what extent are democracy and consumption rational choices?

Social Darwinism and individual choice

Charles Darwin was the pioneer and the genius of his epoch. He discovered that the theory of transmutation of species rested on shaky foundations. Rather he proffered an innovative model that explains evolution as a sum of forces which are confronted in a competition where the fittest organisms survive. According to his thesis, populations evolve over time thanks to a process of natural selection. Darwin argues convincingly that species grow in view of two key factors, fertility and their capacity to adapt to the environment. Since individuals in a population vary in their capacities and skills, this explains why some perish while others prosper. What is more important, those individuals which are less suited to the environment have less probability of survival than others. Although Darwin never intended this theory to be applied beyond the borders of biology, his nephew Sir Francis Galton envisaged the opportunity to continue Darwin's efforts by extrapolating them to human organizations. Society would work along similar lines to those Darwin observed as persons are not equal among themselves. Galton triggered a hot debate respecting the nature of humans and he laid the foundations of a new theoretical wave, social Darwinism. The exegetes and proponents of this theory overtly applauded the idea that some races were smarter and stronger than others, and for that reason, more suitable for monopolizing the means of production of society. Although in Europe and the US, social Darwinism adopted a different connotation, it is no less true that, at heart, it postulated the supremacy of English-speaking society or Norse culture over other ethnical groups. From its outset, social Darwinism was stubbornly rejected by religious authorities as well as Protestantism because it confronted "the sacred text" but gradually and in the US, Darwinism was conceived of as a sign that Protestants were morally superior to other faiths. The laissez-faire society placed all people in egalitarian conditions but, because of constitutive features, each one wanted to outstrip the others towards the formation of an exemplary culture. Likewise, business and capital owners would be the most recognized achievers in a community where individual effort marks the difference between success and failure. In that way, social Darwinism slowly penetrated the circles of religiosity, forging not only American nationalism but also a dangerous ethnocentrism whose zenith was the rise of Adolf Hitler to power (Korstanje 2016).

One of the most critical voices of social Darwinism, Richard Hofstadter, claimed,

> In some respects, the United States during the last three decades of the nineteenth and at the beginning of the twentieth century was the Darwinian country. England gave Darwin to the World, but the United States gave to Darwinism an unusually quick and sympathetic reception.
>
> (Hofstadter 1944: 12)

In perspective, in his book *Social Darwinism in American Thought*, Hofstadter attempts to describe the influence of social Darwinism on the American intelligentsia and politics. The struggle for existence played a vital role in the configuration of a social imaginary which produced specific narratives and discursivities which remain to date. Here Darwinism should be dissected according to two main tenets. On one hand, the idea of natural laws that reigned in Darwin's works would be extrapolated to social sciences. On another, Darwinists praised the competitive order legitimizing the already-existent inter-class inequalities as the consequence of "natural selection". In view of this, the interplay of these ideas with liberal policies for the economy (derived from the ethics of Protestantism) mutated into a radical aversion to state intervention. Government activism was seen as a token of moral decline, which would inevitably lead towards economic bankruptcy. The advance of labor unions adjoined to the rise of a stronger workforce was conceived by Darwinists as the major threat to American peace. As Hofstadter puts it, social Darwinism not only proclaimed the inferiority of some ethnical groups respecting the white man but also generated an irreversible solipsism and individualism which were legitimated in an atmosphere of competition and self-achievement. Toying with the belief that the survival of the fittest was the touchstone of social Darwinism, as Hofstadter maintains, conservative thinking suggested that Darwin's findings in biology can be replicated in industrial civilization as well. Society can be compared to an organism subject to tensions with other organisms and the environment. At the same time that this theory explained why the winners took all, it legally justified the exploitation and exclusion of thousands of citizens who were excluded from the "American way" (Hofstadter 1944).

Lastly, Hofstadter's legacy shows two important aspects of capitalism. At first glance, the would-be climate of liberty is not given to the emancipation of working people but to configure the platform from where they shall compete with others. Further, beyond the veil of democracy, we find a technocratic society which is ruled through the articulation of rationality as the main value.

The society of consumers

Consumption and consumerism seem not to be new or at best they are not monopolized by contemporary society and can be studied in other contexts and times. British historian James Davidson (1997) explores the roots of consumption in Ancient Greece. He conducts an interesting investigation which combines ancient sources with some in-depth anthropological insight. In fact, Greek culture not only left a substantial background in testimonies, books and philosophical treatises but is also situated as the *Cradle of Democracy*. As Davidson puts it, democracy was born in Athens but was not widely practiced in Greece. It consisted in an innovative way of ruling the city. Although it was not a "representative democracy", but a direct one, regulated by an Assembly, democracy inaugurated a system of selection where any member may propose a theme of discussion, proposing laws or rejecting them according to the collective interests. However, this climate of egalitarianism notably contrasted to the world of slaves. In fact, Athens was based on an aristocracy, where the cultural values were centered on the leisure of consumption, banquets and feasts. Consumption in Athens was articulated according to an anthropological gift-exchange ritual which was conducive to the existent money and power asymmetries (Davidson 1997). At the time "the direct democracy" set the pace for a democracy of representants, consumption and leisure expanded an as ideological instrument of control (Veblen 2005).

Doubtless, though the sociology of consumption has gained traction over recent years, and is pivotal in thousands of productions and editorial publications, two authors are of paramount importance in presenting a clear diagnosis of the society of consumers: Zygmunt Bauman and Kathleen Donohue. While the former departs from the fields of philosophy, alternating the theory of rationality with his diagnosis of liquid times, the latter developed a historical diagnosis that explains the passage of a society of producers to a society of consumers. Bauman embraces enthusiastically a caustic critique against capitalism, adjoined to a classic Marxist position, whereas Donohue places Marxism and its fears regarding domination and the rise of poverty as the key factor towards a society of consumers.

Let us clarify for readers: although initially his books and papers are widely cited, Bauman's legacy is very hard to grasp, even for experts in the field. His concerns revolve around the effect of a culture of consumerism and the different shifts industrial society faces. In his terms, the society of production not only exhibits a gap between haves and have-nots but also assumes that the division of labor replicates a system of oppression and exploitation of the workforce. In one of his seminal books, *Modernity and the Holocaust*, Bauman (1989) explains that far from being a direct result of emotionality, the crimes against ethnic minorities and civilians which were perpetrated by Nazi Germany corresponded to the triumph of rationality and the ideals of Enlightenment. The Christian antisemitism that dated from centuries ago, through the invention of rational programs of extermination, was organized according to unprecedented cruelty. Basically, as he discusses, Auschwitz represents an extension of the modern factory, which, instead of producing goods and reprocessing raw materials, exploited bodies to produce death.

> The truth is that every ingredient of the Holocaust – all those many things that rendered it possible – was normal; normal not in the sense of the familiar, of one more specimen in a large class of phenomena long ago described in full, explained and accommodated (on the contrary, the experience of the Holocaust was new and unfamiliar), but in the sense of being fully in keeping with everything we know about our civilization, its guiding spirit.
>
> (Bauman 1989: 9)

This above-mentioned point is vital to understand Bauman's thinking and of course his contributions to social sciences. While the capitalist system of production works successfully, further waste is inevitably produced. In the later capitalism, the yield of wealth, which is concentrated in a few hands, seems to be directly proportional to the rise of poverty. Capitalism not only exploits rank-and-file workers but it eradicates the essential human experience, transforming workers into commodities. In *Consuming Life* he toys with the belief that new tendencies in consumption have changed day-to-day life, overriding the old established customs into new ones. At the same time that companies often eliminate those workers who are considered unfit for production, politics is transforming human resources into commodities. Bauman is correct when he confirms, through the capitalist cosmology, that everything (or at best everyone) that is not efficient should be replaced in its functioning. It is important not to lose sight of the fact that consumers are prone to a self-commoditization of the goods they finally choose. To put this in other terms, even if the market has historically highlighted the encounter between supply and demand, hitherto the boundaries of both are being blurred. Tenets of capitalism are based on attractiveness of elaborated products that can be sold to a wider net of consumers. For Bauman, secularization involves not only religion but also politics. In the past, citizens were interested in public affairs and the claims of politicians for solutions to their concerns. Today, the market seems to have invaded and replaced the role of the State. Rules of the market have been expanded to public life and determine what is or is not due. We should assume that we live in a society of consumers characterized by a lack of neatness between consumer and goods (Bauman 2013). He introduces the term "fetishism of humanity" to denote the sense of security, screens and an ocular-centric character given to lay peoples. Our current obsession for gazing brings some stability to a life of real instability, where the basic needs of workers are always at stake. The metaphor of security, for Bauman, would be an ideological discourse oriented to protect the interests of the ruling elite. The television first and the web now act as conduits that mediate between peoples. Social ties are not only destroyed but the public audiences developed an uncanny fascination with disasters and risky situations. Lastly, as he discusses, in the liquid society of consumers, goods are produced not to endure for a lifetime, but to be constantly replaced. This generates an ever-increasing demand for consumption, which affects somehow the social bondage. A decline of trust and personal contact is accompanied by the quest for novelty. Pointillist time is characterized by the promotion of uncertainty and novelty. Prioritizing the capacity of accumulation, nowadays capitalist societies reward the possibilities of change. Most certainly, workers enter in a climate of hyper competence where emotionality is instrumentalized to domesticate private life. This seems to be a society which is based on rational thinking and value-oriented short-term actions, where *narcissism* plays a leading role in the configuration of an individualist culture (Bauman 2013). The question of whether fear is derived from the decline of social bondage, is a point Bauman adamantly continues (jointly with David Lyon) in *Liquid Surveillance* (Lyon & Bauman 2013).

One of the points that help us to understand his development depends on the intersection of surveillance-related culture with the world of digital technology. Subordinated to the production system, workers not only are commoditized but they are afraid to be excluded from the trade circuit.

They should compete not only to be elected as a product, but also not to be excluded from the formal trade circuits. It is interesting to discuss to what an extent workers do their best to avoid the symbolic death. Even though, both Lyon and Bauman acknowledge that 9/11 did not create in fact the logic of liquid surveillance, it accelerated the conditions of reproduction. Employing the term adiaforization, as a dissociation between action and ethic fields, Bauman adds that the introduction of technology originally was aimed at mitigating some major risks. However, it has paved the pathways for the advent of actions which are not linked to ethics, the subject at some extent, has not developed any commitment with the consequences its action generates. The other was reduced to be subject to the operalization of machines, digital instruments manipulated by automats. Any error, any mistake at time of calculating an attack, should be labeled as "collateral damages".

In this vein, Bauman and Lyon allude to what Arendt called, the banality of evil which means the burocratization of critiques over reason. Although Foucaultian observations were widely employed to study the social behavior in last decades, authors reply that now things have changed a lot. The old panoptic which suggested that few may watch many people, has set the pace to another reality. Few are gazed by the rest of society. A sense of imposed mobility given to all who can pay for that, but at the same time others are immobilized. The archetype of tourists, as ambassadors of their cultures, or capital owners is contrasted to the future of migrants, who are traced, jailed and deported year by year.

(Korstanje 2015)

The 9/11 attack not only accelerated a long-dormant fear in the US, but it liberated a logic of surveillance which goes through two contrasting realities. On one hand, the world of tourists, this first world privileged class which is legally invested to travel across the globe. On another, "an undesired class", usually formed by asylum seekers, migrants and refugees, are often monitored, jailed and expulsed from the wonderland. The culture of surveillance signals to protect the security of the homeland and at the same time the beholders of these technologies feel that they are part of the "chosen peoples". On one hand, it is imperative to control the Other which remains to be an undesired guest. On another, use of technology marks the citizen as a good person. Those who use technologies of surveillance are demonized as "criminals". In doing so, undesired guests who cannot pay for these technologies are marked and pushed to the peripheries of the city.

Lyon and Bauman's efforts suggest two significant ideas, which deserve attention. Firstly, the liberal state is unable to protect its citizens because the power was conferred to trade. The modern nation-state is obliged to give solutions for problems created elsewhere. This engenders a sense of anomie, by which the citizens feel vulnerable. Secondly, the introduction of surveillance technology makes an unsafe world. The quest for order that characterizes the human existence is determined by the needs of change.

To this point, we have reviewed the main aspects of Bauman's thought, revolving around themes such as consumption, security and social order. Kathleen Donohue, unlike Bauman, is not interested in a radical view of capitalism but in a historical description that gives some hints as to how and why the society of producers set the pace to a decentralized economy, which paved the way for the rise of a society of consumers. In this vein, her work, which is entitled *Freedom from Want*, explores the different theories around consumption and the fear of economists that envisaged consumption as a disruptive force tending to chaos and wealth destruction. As she observed, Franklin Delano Roosevelt, in his public speech about the four freedoms (fear, speech, religion and want), mentioned the *freedom from want* while in fact less attention was paid to it by academicians. If economists historically pondered the importance of the division of labor as the centerpiece of production and prosperity, no less true is that the idea of modern consumerism inverted the message.

> Even the classical liberals turned their attention to eradication of poverty; they continued to emphasize production rather than consumption. If one was entitled to consume only what one had produced, then, classical liberal reasoned, the only way that government could eliminate poverty was by increasing productivity.
>
> (Donohue 2003: 4)

With the benefits of hindsight, some radical left-wing scholars held the thesis that capital owners would oppress society to the extent of expanding poverty and the pauperism of the lay workers. The interest of owners was conceived as a negative aspect of society which unless regulated would lead towards greed and self-exhaustion. The frenetic quest for profits led societies to adopt consumer-oriented systems of production that produced what consumers needed. This qualitative view was of paramount importance to understanding the radical change America was internally facing. In doing so Keynesian policies fit like a glove. Donohue argues convincingly that from its inception, *mercantilism* vindicated the idea that the wealth of nations rested on the capacities and skills to trade with other nations. The best way of boosting the economy consists in selling goods to other economies creating a virtuous circle which cyclically would benefit both or more nations. When exports supersede imports the economy rises, in the same way sustainable development would only be possible with a controlled level of internal consumption. However, after the 1940s, the freedom from want was posited as one of the human needs which gradually shifted the background of economic theory. Intellectuals henceforth began to discuss the leading role played by consumerism to reduce poverty as well as laying the foundations to compare consumption with democracy. This new liberal system not only faced one of the most terrible stocks and market crises of capitalism in the 1930s, but also contributed to the expansion of a radical view which claimed that capitalism was responsible for the hapless situation of millions of American citizens. Detractors of capitalism, who pushed their focus on the arbitrariness of producers, were involuntarily responsible or conducive to the formation of a global society of consumers. Those denunciations of an economy that protects the interests of producers as well as the need to adopt consumption to break the material asymmetries among classes were two guiding concepts to embrace a globalized version of capitalism, prone to mass consumption. In this respect, the society of consumers paradoxically applauded democracy (as the best form of government), which encourages liberty and egalitarianism but at the same time, consumerism neglects individual rights, enslaving citizens to succumb to their own passions (Donohue 2003).

Both writers, from different angles, have systematically thematized liberality, consumption and democracy. While Bauman, who was notably influenced by Marxist theory, focused on philosophical materialism, Donohue, rather, goes on to an alternative argumentation, which rested on a historical review of liberalism. The Marxist economists developed an uncanny fear of poverty, which resulted in the adoption of consumption as a positive value. The sense of ownership has been demonized without rational basis to the extent of leading society to the arbitrariness of hyper-consumerism.

References

Amadae, S. M. (2003). *Rationalizing Capitalist Democracy: The Cold War Origins of Rational Choice Liberalism*. Chicago, University of Chicago Press.

Bauman, Z. (1989). *Modernity and the Holocaust*. Ithaca, Cornell University Press.

Bauman, Z. (2013). *Consuming Life*. New York, John Wiley & Sons.

Bernstein, R. (1988). "The Rage against Reason". In *Construction and Constraint: The Shaping of Scientific Rationality*, E. McMullin (ed.). Indiana, Notre Dame University Press, 189–223.

Blais, A. (2000). *To Vote or Not to Vote? The Merits and Limits of Rational Choice Theory*. Pittsburgh, University of Pittsburgh Press.

Davidson, J. N. (1997). *Courtesans & Fishcakes: The Consuming Passions of Classical Athens*. New York, St. Martin's Press.

Donohue, K. G. (2003). *Freedom from Want: American Liberalism and the Idea of the Consumer*. Baltimore, Johns Hopkins University Press.

Downs, A. (1957). *An Economic Theory of Democracy*. New York, HarperCollins.

Dryzek, J. S., Honig, B., & Phillips, A. (Eds). (2008). "Introduction". In *The Oxford Handbook of Political Theory* (Vol. 1). Oxford University Press, 3–43.

Feyerabend, P. (1982). *Science in a Free Society*. London, Verso.

Feyerabend, P. (2012). "How to Defend Society against Science". In *Philosophy: Basic Readings*, N. Warburton (ed.). Abingdon, Routledge, 261–271.

Hobsbawm, E. J. (1995). *The Age of Extremes: A History of the World, 1914–1991*. New York: Pantheon Books.

Hofstadter, R. (1944). *Social Darwinism in American Thought* (Vol. 16). Boston, Beacon Press.

Ingold, T. (2000). *The Perception of the Environment: Essays on Livelihood, Dwelling and Skill*. London, Psychology Press.

Klein, N. (2007). *The Shock Doctrine: The Rise of Disaster Capitalism*. New York, Macmillan.

Korstanje, M. (2015). Book review of *Liquid Surveillance* by Zygmunt Bauman and David Lyon. *International Journal of Baudrillard Studies, 12*(1).

Korstanje, M. (2016). *The Rise of Thana-Capitalism and Tourism*. Abingdon, Routledge.

Lyon, D., & Bauman, Z. (2013). *Liquid Surveillance: A Conversation*. Cambridge, Polity Press.

Ostrom, E. (1998). A behavioral approach to the rational choice theory of collective action: Presidential address, American Political Science Association, 1997. *American Political Science Review, 92*(1), 1–22.

Pells, R. H. (1989). *The Liberal Mind in a Conservative Age*. Hanover, University Press of New England.

Schlosberg, D. (2008). "The Pluralist Imagination". In *The Oxford Handbook of Political Theory*, J. S. Dryzek, B. Honig, & A. Phillips (eds). Oxford University Press, 142–162.

Skoll, G. (2014). *Dialectics in Social Thought*. New York, Palgrave Macmillan.

Stilz, A. (2009). *Liberal Loyalty: Freedom, Obligation & the State*. Princeton, Princeton University Press.

Veblen, T. (2005). *The Theory of the Leisure Class: An Economic Study of Institutions*. State College, The Pennsylvania State University.

Virilio, P. (2010). *The University of Disaster*. Cambridge, Polity Press.

Weingast, B. R. (1979). A rational choice perspective on congressional norms. *American Journal of Political Science*, *1*, 245–262.

3 The rise of terror in the society of the spectacle

Introduction

Recently, a news source indicated that military camps in Israel were luring foreign and national tourists. These "terror camps", geographically scattered through Israel and the West Bank, offered a morbid spectacle that in fact evinced some risk-taking on the part of the tourists. Among the activities, tourists had the chance to perform the role of IDF soldiers in some spectacularized situations. The costs range from USD 100 to USD 1000, including a lot of activities where visitors emulate the tasks of Israeli soldiers in the detainees' camp. This event led me to think about some questions which deserve to be discussed, such as, why are these types of morbid forms of tourism emerging? Is war-tourism or terror-tourism a new form of empathy with the Other? And, of course, why do first-world tourists need these radical experiences?

The first section of this chapter deals with the question of entertainment, the radical interrogations on the place of the alterity as well as the conceptual discussion revolving around the discovery. Of course, the sociology of spectacle has advanced considerable steps in the understanding of why we need to gaze at the Other's pain. By this token, the second section explores the uncanny intersection of terrorism and tourism. Here the voices are divided. While some experts allude to the fascination showed by the media for the coverage of terrorism-related news, others alert us to the risk the society of spectacle offers in mediating "morbid forms of consumptions" such as war-tourism and "dark tourism" as valid forms of entertainment. This is exactly what happens in sites plagued by mass death, genocides and traumatic experiences; these dark sites are often – like these military camps – visited daily by thousands of tourists. We propose a new theory to understand this reality. The society of risk sets the pace to a new facet of capitalism, which we dubbed as Thana-Capitalism and where the Other's death plays a leading role. To put this paradox bluntly, our parents visited paradisiacal landscapes or enjoyed their days at beaches. Now, tourists prefer darker products such as visits to concentration camps, military camps and spaces of mass obliteration such as in New Orleans or Sri Lanka. Doubtless, these new types of tourism coincide with the rise of a new society, which is based on the need to venerate death as the main cultural value.

The sociology of spectacle (explained with clarity)

Guy Debord was one of the pioneers who envisaged the role of spectacle in contemporary society. According to his original diagnosis, human activity as it was
known by ancient cultures was prone to suffer substantial shifts. The fetishism
of the commodity vaticinated by Marx and Marxian scholars assumed the inver-
sion of values, which remained in "congealed forms", into abstract commodities
which are exchanged through the lens of the spectacle (Debord 2012). In consonance with this, Norbert Elias and Eric Dunning (1986) investigated the essence
of sports and leisure consumption as the natural quest of excitement modern citizens needed to escape the humdrum routine. The sense of a controlled risk is the
touchstone of modern capitalism, which is oriented to depersonalize emotions.
Lay-people in the public sphere are unable to express their emotions unless at
sporting spectacles. With the same orientation, Dean MacCannell claimed that
capitalist society can be very well compared to the tribal culture, in a way that
– at best – other sociologists have glossed over. The totem emanates a source
of authority and symbolism which legitimates the ruling elite as well as other
sacred rituals. As an instrument of religiosity and control, the totem organizes the
cosmology of aboriginal cultures and the ways they are adapted to the environment. To some extent, this was applied by Ancient Europe but once the process of
secularization was introduced to modern society, religion inevitably declined.
There was a gap, which was fulfilled by modern leisure and tourism. Hence for
MacCannell, tourism plays a leading role in avoiding the rupture of society –
like the totem in the aboriginal tribe (MacCannell 1976, 1992, 2001). Under the
notable influence of structuralism, his thesis is that societies move according to
a much broader cultural matrix. Social scientists should explore the formation
of these matrixes to know how the society works. In this respect, tourism not
only is the totem by other means but molds the borders of a much deeper cultural
matrix which organizes the contemporary means of production. This was exactly
the point of discrepancy between Dean MacCannell and John Urry. For the latter,
one of the troubling aspects of modernity is the ambiguous dynamic of mobilities,
which may generate a free movement in one direction with the strictest restrictions
in the other. As he puts it, the car seems to be the best example of what he dubbed
the paradoxes of mobilities. It fostered an extension of the proper body at the same
time as it exacerbated the quest of an emancipatory discovery (Urry 2004). To
some extent, the powers of eyes are liberated – through the tourist gaze – but with
the risks of coercing people into inflexibility. Since the introduction of modernity
has diluted the centralized means of production that characterized other stages of
humankind, as Urry wrote jointly with Scott Lash, not surprisingly the essence of
commodities have taken on a more abstract form. Capitalism not only expands
through the economy of desire (gazing) but also started with a character that pri-
oritizes the signs of products over other forms of consumption. In view of this, we
face the end of "organized capitalism" into more decentralized forms of production that paved the way for the rise of a circular (aesthetic) reflexibility. For Urry
the society of which Marx and his followers dreamed is obsolete, simply because

capitalism encouraged a faster (more mobile) mode of exchange where cultures, peoples and landscapes are commoditized (Lash & Urry 1992). In this context, his theory of mobilities and, above all, the concept of tourist gaze play a leading role.

In 2011, Urry published (with J. Larsen) an updated version of his classic book, *The Tourist Gaze*. In this version, doubtless, the concept of gazing, circularity and spectacle deserve our attention. Like MacCannell, Urry realizes that tourism plays an important role in generating pleasurable experiences. People often crave, hope and move to far-away places to grasp "experiences" that otherwise would be untouched. The gaze not only mediates between tourists and their landscapes but also organizes what should be gazed upon. Of course, each gaze looks at something different – the medical gaze and the tourist gaze are subject to different backdrops – as he adds.

> Gazing at particular sights is conditioned by personal experiences and memories and framed by the rules and styles, as well as by circulating images and texts of this and other places. Such frames are critical resources, techniques, cultural lenses that potentially enable tourists to see the physical form and material spaces before their eyes as interesting, good or beautiful.
>
> (Urry & Larsen 2011: 2)

The above extract, for the sake of clarity, exhibits Urry's main concerns, which legitimate the attention of tourists for some landscapes, while others are discarded. In sharp contrast with the notion of authenticity for MacCannell, which received a pejorative definition as alienation or confiscatory force, the tourist gaze (no matter how light or dark it may be) seems to be individually acquired, alternating not only the ideological discourses imposed by the ruling elite, but also the individual and psychological worlds of tourists. The act of watching involves possession, and modern consumers are excited to possess what they do not have; but all these practices, as Urry states, are certainly conditioned by the culture. The cultural matrix that MacCannell denounced as an ideological apparatus appears to be a result of the economic organization which is based on mobilities, spectacularity and aesthetic consumption (Urry & Larsen 2011). Though the "darkness" does not occupy an important space in his argument, he explains the September 11 attack as the interplay of contrasting values such as complexity and the networked globalization, the micro and the macro levels, as well as the dialogue between wild and safe places. While the risk invests a net of experts to sort the world according to Western rationality, the aesthetics of capitalism opens the doors of a circular logic of uncertainness that divides the world in two: safe (civilized) and wild (relegated) zones (Urry 2002). The efforts of intervening rationally into the wild area instill further panic and anxiety as Urry brilliantly concludes.

As noted, this point was addressed by R. Tzanelli in her book, *Thanatourism and Cinematic Representations of Risk*. As she observed, there is an emerging chasm between sightseers and the gazed-upon natives. Beyond the pleasure of gazing at dark events such as the terrorist attacks on tourist destinations or even natural disasters, a concatenation of risks and representational discourse revolving around the risk are being broadcasted by cinema and media. Of course, Tzanelli asks, why? Under the auspices of the spectacle logic, as she discusses, dark consumption – through the lens of the exploited "Other" – acts as a mechanism of surveillance to keep the citizens docile. It is important not to lose sight of the fact that cinema offers two important things. On one hand, it packages and disseminates a distorted explanation of disasters, which is culturally oriented to embrace the proper Western heroes, but on another, the landscapes of the end of days – far from correcting the problems that may lead society to disaster – replicate the ground to protect the interests of the status quo. To put this bluntly, though Europeans visit long distances to be in contact with the poor of Rwanda and the sad experience with genocides, their States never accept liability for the cruelty when colonial rule prevailed. In that way, Tzanelli argues convincingly, dark cinema today offers the possibility to introduce an ideological discourse dislocating the historical contexts, where events occur, from a fabricated story of the exploited "Other" (Tzanelli 2016). In the next section, we shall discuss the intersection of (dark) consumption, terror and attraction, a point which helps us to better grasp the nature of post-disaster destinations.

Tourism in the days of terror

The 9/11 attack inaugurated, for some voices, a new era of uncertainty and terror not only on US soil but also throughout the world. Most likely, its frightful nature does not rest in the number of victims or the material losses but in the fact that the means of transport, which were the pride of the Occident, were weaponized against civilian (symbolic) targets (Skoll 2007; Howie 2012; Korstanje 2018). Luke Howie (2011) developed an interesting model which engages terrorism with media attention. Terrorists should not necessarily be considered "sadist" killers, though they commit horrendous crimes. The logic of terrorism corresponds to the culture of celebrities, where costs and gains are rationally evaluated. In some respect, terrorists do not want a lot of people killed, they want a lot of people watching(!), Howie confirms. The culture of witnessing (terror) was functional to both sides. Terrorists introduce their extortion while the media gains further publicity for investors. Taking his cue from Baudrillard's contributions, Howie explains that one of the paradoxes of terrorism relates to the fear awoken in remote, unaffected countries, like Australia. It is safe to say that those nations or people who are apart from the targets or the operational hub of terrorism are the most terrified. This happens, according to Howie, because,

> The spectacle of terrorism depends on the co-existence of witnesses, images of terrorism, and – in contemporary times – cities. 9/11 happened, it happened on 11 September 2001 in New York City, Washington DC and a field of Pennsylvania. The image, however, is not bound to this temporal and geographic logic. 9/11 was an atemporal event that can be understood in time

and space in apparently unlimited coordinates of temporality and spatiality. It resides in the desert of the real of the contemporary city.

(Howie 2011: 60)

He coins the term, phenomenology of terrorism, by reflecting how modern consumption emulates a frightening climate, which serves paradoxically as an instrument of entertainment. The power of amplification the mediated terrorism offers captivates global audiences while the media gain further profits. This constitutes a culture of witnessing which is designed to conduit terror as the blood the capitalist veins need. Undoubtedly, academicians of all stripes agree 9/11 was a founding event that accelerated the logic of entertainment, which was already enrooted in the popular culture. David Altheide (2017) brings reflection on the advance of ISIS, the terrorism-related news and the radical discourse of Donald Trump against mass-migrations as issues, which are part of the same phenomenon. The institutions for checks and balances not only cede when the paralyzing force of terror advances, but individual rights which are legally granted by the constitution often stand back at the current political stage or are completely sidelined. Trump not only is a product of terrorism but is from a society which requires the spectacle of terrorism to feel special (Altheide 2017; Korstanje 2016).

Given the discussion in these terms, another interesting question lies in knowing why terrorists target tourist destinations to achieve these cruel goals.

Popular opinion precludes the possibility that terrorists are frustrated persons who are motivated emotionally. Their hate is directed against vulnerable tourists in the same way their worlds scorn the development of Western civilization. This so-called clash of civilizations (paraphrasing Huntington) would be the result of the triumph of democracy over other cultural values. Tourists are ambassadors of their own (civilized) nation and for that they are unjustly the reason for the violence (Sönmez & Graefe 1998; Sönmez 1998; Pizam & Smith 2000). However, Professors Enders and Sandler (2011) show empirically the opposite. Terrorists – like any rational agents – look to maximize the gains while their costs are minimized. In a hyper-globalized culture where news arrives for the entire globe in seconds, terrorists need to spend little energy in disseminating their crimes. Meanwhile, they ensure devastating effects on the public because they vulnerate public spaces. The message seems to be simple, the same can happen again anytime and anywhere, there is nothing your State can accomplish to protect you! In this way, they look to undermine the legitimacy of States in order for the terrorists' demands to be obeyed. Tourism is particularly sensitive to terrorist attacks and this is still the object of study of some colleagues and experts. Raoul Bianchi (2006) called attention to the expanse of fear, as something other than a logic of domination, but as the apotheosis (the foreclosure) of consumer capitalism. The world is based on clear paradoxes. While free movement signals the triumph of democracy and the harmony of capital owners and the workforce, political instability, crises and violence arise on both sides of the river, in the first and third worlds. Terrorism has historically incited military intervention, in which case, the dilemma of security becomes the main worry of the West. To put this in other terms, believing in the "right to travel" entails that we accept there are spaces of conflict and chaos which may place our ontologies in jeopardy. In a recent book, *Tourism and Citizenship*, Bianchi and Stephenson (2014) exert a radical criticism of the neoliberal discourse, which aims to outline the role of globalization in the pacification of the world. Tourism is never affected by terrorism, it is an ideological construction intended to reinforce the center–periphery dependency. The introduction of luxury consumption leads us to think that tourism is the only vehicle towards development. However, the recent growth of tourism is not reflecting a more egalitarian world. Poor nations are dominated, following Bianchi and Stephenson, by rich economies. The process of globalization often interrogates the Third World, recalling its so-called inferiority with respect to the developed world. Tourism should be understood by these authors as something more complex than an industry but as an instrument that strengthens the material asymmetries between capital owners and the rank-and-file workers. This begs a more important question: to what extent may we see tourism as part of the problem or part of the solution?

Abundant evidence suggests that tourism can accelerate the time to recovery in post-disaster contexts in a way detractors of tourism never imagined (Biran et al 2014; Mair, Ritchie & Walters 2016; Walters & Mair 2012; Séraphin, Butcher & Korstanje 2017). In recent years, post-disaster or post-conflict marketing offered excellent conditions to assist locals to recover, strengthening the resiliency of society (Shondell Miller 2008).

Dark tourism, post-conflict destinations and mourning in late capitalism

A critical position around post-conflict destinations was excellently described by Comaroff and Comaroff, who alerted us to the risk of heritage consumption in lands of previous hostility and conflict. Without doubt, ethnic tourism or heritage tourism has globally contributed to the improvement of many oppressed tribes and cultures. Through the articulation of tourist destinations, casinos and leisure sites, many relegated tribes escaped from a programmed death. No less important is that tourism revitalized the economic asymmetries in societies oppressed by colonialism or any other type of undemocratic government. However, as these authors lament, in some conditions the nation-state, eager to amass further resources, imposes heavier taxes on the growing new projects. Analysts should understand that, far from solving the historical ethnical disputes, when tourism is not regulated such disputes are aggravated (Comaroff & Comaroff 2009). Then, as Comaroff and Comaroff remind us, specialists should weigh the costs and benefits of tourism consumption. Another important aspect of this slippery matter indicates that disasters cannot be evaluated or judged by their immediate effects, but through the ways they are internalized by survivors (Keane 2006). At the time disasters hit two major problems emerge. On one hand, the established protocols, which were originally designed to be followed in case of disasters, are overlooked in view of the climate of panic that often persists. On the other hand, disasters not only leave a lesson to the community but appeal to the formation of ideological discourses, which are historically and geographically embedded with social practices (Lacy 2001). Dark tourism as an emerging sub-discipline dissects the formation of discourses revolving around the social imaginary that sells sites of mass destruction as products (Johnston 2013). As enrooted in a specific territory or soil, dark-tourism practices are encapsulated in forms of heritage, which offers an interpretation to make the trauma more human (Poria 2007; Raine 2013). Erik Cohen (2011) adds the term populo-site to define the contours of the place where the memory is founded. While there are a lot of sites which emulate the memorized event, only a few provide a real experience for tourists. This is the place where everything started, where the bodies fell for the first time, or even where the founding heroes were buried. Cohen maintains that dark tourism serves as a pedagogical mechanism of learning for a similarly minded event to be avoided in a not-so-distant future.

Still further, White & Frew continue Cohen's worries, commenting that dark consuming sites are politically designed to send a message to a community, but far from being unilateral, such a lesson is adamantly negotiated by each member. The fact seems to be that there are no clear-cut borders to understand why and when a site offers a unified or accepted lesson while in other terms there are scattered and fragmented interpretations revolving around the same event. While heritage relates to political interests, in some contexts the national discourse may operate on previous ethnocentric discourse or ignite radical chauvinist policies against some ethnic minorities (White & Frew 2013).

In *Heritage that Hurts,* Joy Sather-Wagstaff (2016) confronts the notion of dark heritage, at least as it is presented in these types of traumatic events. By this token, there is a sentiment of brotherhood which puts all people in egalitarian conditions before death. She holds that this sentiment of reciprocity, which seems to be the touchstone of resiliency, helps the community to keep united while the lessons run serious risks of being politically manipulated by the status quo. The essence of heritage corresponds to a distortion of the lesson left by the event, a politically designed discourse, so to speak, that imposes a biased explanation of why the disaster happened. In fact, Sather-Wagstaff explores the risks of dark heritage while alerting us to the problems and limitations when professional politics take direct action.

Lastly, as Korstanje (2016) puts it, the risks of distorting the real facts behind the disaster are higher, because society is doomed to replicate the basis that led to such a tragic event. In a society that valorizes the Other's pain to affirm the status of viewers, imposing an *ethical remembering* is one of the challenges of modern consumers. From torture to death, Western civilization has fleshed out an uncanny fascination for death which merits discussion below.

Consuming death

When I developed the idea of Thana-Capitalism, I thought of the impossibility of sociology, or at least the postmodern sociological lens, to provide concrete indicators that help me to read the phenomenon of dark tourism. I found, even, that under the tip of this iceberg there was a profound anthropological explanation as to why we are culturally obsessed to gaze at the Other's death (Korstanje 2016). As P. Stone puts it, dark tourism evinces something other than the visitors' interests in mass death; it represents an anthropological attempt to interpret one's own life through the Other's death (Stone & Sharpley 2008; Stone 2012). Today's tourists understand the importance of "being there" – like the first ethnographers – at the time things happen. At a closer look, tourists go to the war to see the performance of armies or simply employ technology to enter sites that otherwise would be impossible (Kaelber 2007; Germann Molz 2012; Korstanje & George 2017). As noted, the term *Thana-Capitalism* has its etymological roots in Thanatos, which means death. My own conception of Thana-Capitalism aims to indicate that the consumption of death seems to be everywhere, not only in tourism but in contemporary society. We watch how others die in the TV magazine, in films, documentaries and video games among other forms of entertainment. In the 1940s, one of the fathers of symbolic interactionism, George H. Mead (1934), argued that the conception of the self is constituted in relation to the alter ego. Although people were accustomed to complain about the bad news in the newspapers, Mead adds, they can't stop reading them. This happens for two main reasons. Firstly, the relation between each self is based on its interactions with others. Secondly, the self experiences a sentiment of happiness when the Other is harmed or dies in view of the fact that the self avoided being touched by the death.

Paradoxically, the press plays an ideological role domesticating not only the future, but also imposing the conditions under which the citizen feels its world is the best of all possible options. No matter how we are, no matter what we do, the Other is always suffering and this suffices for us to accept democracy as the best form of government. A second point of entry to be introduced in this discussion is the theory of narcissism as formulated by Christopher Lasch. In the book *The Culture of Narcissism* (1991) he exerts a radical criticism of American society and its lifestyle. The exacerbated consumerism, adjoined to individualism, not only wreaks havoc in the social scaffolding but also creates a climate of extreme competence where citizens feel special and invested with the right to interact only with others who are special like them. This created a super-auratic caste (of supermen) who are not permeable with the surrounding environment. The culture of narcissism lacks ethics, because it avoids any type of empathy with the Other. The psychological frustration of the ego impedes a fluid dialogue with the external world and the genuine other, but the world is constructed according to internal projections, hopes and needs. Hence citizens in the days of Thana-Capitalism are prone to pleasure maximization no matter the costs. In this respect, the other element, which was widely discussed in earlier chapters, is the rise of social Darwinism as the most vivid expression of this narcissism that claims "the birth of a superman". Not surprisingly, the ideals of Darwinism and the survival of the fittest were cultivated in Nazi Germany as well as in the United States. The Nazis orchestrated a death-machine that instrumentalized the rational bureaucracy to implement a mass death for those who were catalogued as "unfit", "inferior" or "sub-humans". They secretly not only perpetrated appalling crimes against civilians but also inaugurated the meaning of "creative destruction". On the other side of the ocean, Americans fostered their own allegories respecting Anglo supremacy, demonizing Nazis but adopting the same ideology. Finally Hitler was defeated but social Darwinism won the war. As I explained, the logic of the extermination comes from the mythical foundation of the world, once – as the Bible says – Noah receives the mandates of God. This was not only the first genocide but also the first selective destruction God has planned by humankind. As in all myths, these stories speak of the cultural values of the societies behind them. God asks Noah to construct an ark, selecting a pair of each species. The mankind which was corrupted by greed and sin should be destroyed in order for a new life to flourish. Noah not only abided by the wish of the Lord but does not tell others about their grim future. This is the first time the world is divided in two, victims and witnesses. The crucifixion of Christ is not a founding event but it reinforces the same logic. What is more important, if the process of secularization pressed religiosity towards the periphery, why does religion influence daily life or the configuration of social Darwinism?

This secularization introduced a new discourse to day-to-day life. The world of God is not important for men, and of course, death is not the start of a new life. In medieval times, peasants bore a life of hard deprivations knowing that once they are in heaven they will be eternally compensated. As a result of death, the idea of the after-life played a vital role in the control of persons. The process of secularization not only questioned the monopoly of the Church and Christianity to explain the world but also liberated death from its constraints. As P. Aries brilliantly observed, in the Middle Ages death was everywhere and people had a life expectancy of no more than thirty years. However, death was never for medieval thinking a problem or an object of fear. Secularization accelerated the decline of religion breaking the bridge between earth and heaven. In consequence, the efforts of modern humans were placed to preserve and extend life as never before. The modern cosmology developed a pejorative view of death and paradoxically as Ariès (1975) wrote, liberated death from its cage. In this contemporary society, though the expectancy of life is certainly enlarged, death has become a cause of panic for lay-citizens. Metaphorically speaking, life is imagined as a long race where the winner takes all. This point was particularly examined by Weber and other sociologists, but in Thana-Capitalism a few have much while the rest die with nothing. The discourses revolving around selective evolution as well as the happiness for the Other's death ideologically coincide with the asymmetries produced by society and its means of production. The economy of capitalism rests on serious asymmetries and injustices created by what Marx dubbed as "the commodity fetishism" (Pietz 1985). It gradually reproduced a gap between haves and have nots which is ideologically legitimated by the introduction of different allegories and stories. In order for the workforce not to confront capital-owners over their amassed wealth, the ideology dissuades the former through diverse ways which range from religion to consumption. Fundamentally, the ruling elite preserves its privileges, subordinating the consciousness of workers to their own elaborated discourse. As a result of this, death and life become ideological preconceptions that mark the difference between the doomed and the saved. Those who die are not only marked by fate, but are also considered "inferior" with respect to others who still live. This is the reason why, as Mead found, modern consumers are prone to gaze at the Other's death. It is safe to say that Thana-Capitalism engendered a new class, which I named "death seekers"; they have common features which are shown below.

- Death seekers are moved to discuss events but on rare occasions they are directly committed.
- Thana-Capitalism starts with 9/11 and the spectacularization of disaster headed by CNN and other news media corporations.
- Their pejorative viewpoint of the world coincides in praising capitalism and democracy as the best options.
- They have fleshed out a pathological way to understand death.
- They are narcissistic and often feel superior to others.
- They behold a frightful type of personality which considers the world a dangerous place.
- They love the mythical conflagrations as the fight between evil and good.
- Their pleasure maximization is not given by direct achievement but by the failures (death) of others.

The best metaphor that synthetizes death seekers is the dystopian novel, *The Hunger Games*. The plot takes place in a futurist world almost devastated by the oppression of President Snow, who annually manages a game in which the thirteen districts deliver representatives to fight with others in order to survive. These warriors – like death seekers – celebrate enthusiastically the Other's death simply because they have a higher probability of being the winner. This stopped participants from cooperating with others, subordinating their expectations to Snow's desires. The exploitation is only possible, this novel shows, because the oppressed succumb to the ideologies of the oppressors. Of course, the selective survival, the death of almost all participants and the ideals of the winner-takes-all, are the main values of *The Hunger Games* which can be replicated in other novels or reality shows such as *Big Brother*. The macabre spectacles and simulations in West Bank military camps are part of the same cultural ethos. In the society of Thana-Capitalism the Other's death is the touchstone of all human relations, and the obsession for all citizens. Of course, in this context, terrorism fits like a ring on the finger. Terrorists perpetrate their attacks in high-profile tourist centers while the public audiences enjoy the dark spectacle, reminding them that after all the mourning and destruction they still remain "outstanding", part of the "chosen peoples". Lastly, the *Rise of Thana-Capitalism and Tourism* was positively received by scholars although, being honest, it was widely criticized too. One of the most constructive but no less critical commentaries comes from "Bios and Thanatos in Leisure" by Rodanthi Tzanelli (2017). The main aspects of her objection to the thesis of Thana-Capitalism will be detailed next, as well as my replies to her questions.

Criticism of the rise of Thana-Capitalism

At first glimpse, the theory of Thana-Capitalism reminds us that the risk society was replaced by a new stage of production, where death plays a leading role as the main exchanged commodity. In this respect, she departs from the belief that disasters and apocalyptic landscapes are the dominant part of modern consumption styles, displacing other classic forms of leisure – such as sun and sand – to a peripheral position.

In my book *The Rise of Thana-Capitalism and Tourism*, as Tzanelli admits, there is an obsession for gazing at the Other's death, which confirms the climate of a new solipsist atmosphere of self-affirmation. As she puts it,

> This process permeates the logic of contemporary "culture industries", including those of media (such as film, satellite TV and the internet) and tourism, and oils the capitalist machine's profit-making to the detriment of the poor, the weak and the disenfranchised. These populations are easily positioned as objects of scrutinisation, with "First-World" and affluent consumers as spectators of the drama of poverty, or natural and man-made disasters.
>
> (Tzanelli 2017: 269)

The term Thana-Capitalism denotes something other than a curiosity – if not empathy – for the alterity; it connotes much deeper forms of ideological domination, which express the inequalities that capitalism often generates. Tourism would not be a mechanism of control but a ritualistic performance that expresses and represents the mainstream cultural values of society. In a globalized economy, the suffering which is a result of such a globalization is commoditized as a form of entertainment, production and consumption. In sum, Tzanelli acknowledges the importance of producing further critical debates on dark tourism, but reminds us that solidarities are never fully destroyed by capitalism. She makes an in-depth review of my work, chapter by chapter, focusing on the most important ideas and argumentations. Basically, Tzanelli says,

> Korstanje alternatively argues that Christ's originary suffering split the world into victims and observers, thus laying the foundations of contemporary capitalist exploitation. The suffering Christ is now reproduced by cultural industries as a spectacle in a variety of consumption sites, including those of media and tourism, and Korstanje is quick to provide readings of specific popular cultural products through this lens.
>
> (Tzanelli 2017: 270)

Against the previous backdrop, Tzanelli criticizes the argument (of certain elitism) exaggerating – if not romanticizing – the relations between lower and higher classes. According to her viewpoint, there are some conceptual simplifications at the time of declaring that the ruling elite exerts a radical oppression over the workforce. This corresponds with an old, dormant prejudice which may very well be traced back to the 1970s, under the auspices of radical Marxism. She contends that tourists are daily moved by a wide range of motivations which cannot be homogenized or standardized. Not all media industries are oriented to maximize profits (Korstanje 2016). Hence the idea of Thana-Capitalism can be applied to a portion of dark-site travelers but not to the universe at all. The theory of Thana-Capitalism rests on a perverse fallacy, because it assumes (without any firm ground) that dark or slum-related sites constitute "dark memories" which revitalize the center–periphery dependency that preceded in the colonial period. This stance ideologically validates an ethnocentric assumption which posits European travelers as the Grand Tourists who can be "colonial administrators" nourishing the allegories of the "noble savage". Tzanelli writes that we cannot understand capitalism by adopting the thesis that suffering hosts are dominated by the gaze of the rich guests while host–guest encounters are based on a process of reflexibility where both parties coexist. In this vein, we would not overlook the fact that dark consumption – under some conditions – helps locals to reverse their economic conditions of existence. This is the reason why it is very hard to blame the developed world for all illness, prerogatives and economic problems which historically whipped the under-developed world.

My reply to Tzanelli's criticism

Firstly, I do express my gratitude to Tzanelli for her in-depth review but before continuing I must give further details on some important aspects that she inadvertently glossed over. As a brilliant thinker, she synthetized part of the sophistication of my own theories and the open points to follow in future approaches. The theory of Thana-Capitalism was originally created to describe an increasing trend. I have never mentioned that all society moves towards the goals of "death seekers", the emergent class derived from morbid consumption, as Tzanelli says. Of course, one might confess, tourists (no matter the class or national identity) travel for different motives. However, no less true is that dark consumption, which is empowered in new consumption lifestyles as dark tourism, war tourism and slum tourism, is rising. To put this in other terms, our grandparents would never imagine traveling through the dark soil of a concentration camp, or gazing at a community completely devastated by a natural disaster. By this token, dark tourism is not an exception but only the tip of the iceberg. Society keeps similar practices and trends in other media sources such as video games, films, novels, newspapers, news broadcasting and documentaries. I shall enumerate various examples here which support my point, but firstly explain that the key conceptual component of my theory seems not to be "death's consumption", but "social Darwinism". Sir Francis Galton imagined a society sorted by racial hierarchies which crystallized – as we have seen above – the triumph of rational thinking. Europeans were catalogued as superior to other "races" because they developed new innovations and achievements that transformed their culture. Eugenics marked not only the needs of survival as the main criterion of exemplarity but assumed the unfit agents should perish. Implicitly or not, this is the main ideological core of social Darwinism which is adjoined to the survival of the fittest (strongest). Like-wise, English-speaking citizens, occupying the top of a (imagined) pyramid, were invested to administrate the world. Needless to say, this ethnocentric discourse legitimated the Westernization of the globe, expanding conquest, plunder and domination to autonomous cultures.

Here a pertinent question arises: is this logic persistent in the globalized (trans-cultural) world that Tzanelli writes about?

The answer is, yes. The logic of Thana-Capitalism as well as the obsession for dark tourism, was originally encapsulated in the UK, remaining almost unknown for Spanish-speaking nations. This does not mean that Britons are sadists, but they are particularly educated in a society where Puritanism and the Reforma-tion played a leading role in the configuration of a culture of achievement. As the Weberian studies amply show, the Reformation implanted a new cosmology (which is based on the Book of Life) where God knows beforehand who would be saved and doomed. There is nothing that people can do to enter Paradise, it is predetermined by the Lord's wish. This created a culture of manifest asymmetries where a few have a lot while others are pressed to survive. The liberal thought permeates throughout the UK, not only paving the way for the rise of industrial-ism but legitimating the grim future of workers as in their own hands. The world of medieval peasants gradually witnessed a radical transformation where people were liberated from their familiar attachment. While this mobility pushed thou-sands of peasants to sell their labor, the sense of mobility and freedom was intro-duced as a form of legitimating the new status quo. Marx was right in citing the paradox of slavery. In ancient and medieval times, slaves were pressed to work to their death while paradoxically their living conditions were entirely met. They had the necessary food and clothes to work as a property of their master. Capitalism not only liberated these ties, emancipating these slaves as free citizens but inter-nalized the sense of contingency in order for these new workers to be managers of their own destiny. The ethical quandary was that while freer than their ancestors, hunger – as a form of rational scarcity – configured the material asymmetries between the classes. Therefore, in its constituency, capitalism is a system that produces inequalities and contrasting landscapes among the involved agents.

The process of reflexibility does not make people more equal, but ignites new forms of censorship. As stated, postmodern readers envisaged the idea of reflex-ibility to express their concerns respecting how information was produced and disseminated. The nets of experts, which occupied a central position in industrial society, set the pace for more flexible forms of socialization where lay-citizens gather further levels of information to the extent of confronting academicians. In this society, not only was scientific knowledge affordable to lay people but also information circulated everywhere. It is important not to lose sight of the fact that reflexibility never created more equal institutions, but introduced a more subtle – but no less efficient – form of control. In medieval days, authors and their books were energetically coerced by the violence of nobility and Catholic Church. Nowadays, censorship goes in the opposite direction, offering multiple sources and thousands of studies and books, which mislead even the most erudite readers.

If I type "Marx" into an internet search engine I will obtain millions of results, which are almost impossible to digest. This overproduction gave lay-citizens more accessibility to the information but less understanding of what they daily read.

The theory of Thana-Capitalism exhibits the above-mentioned tendency, expanding so-called standardization which pits worker against worker. In this respect, the metaphor of *The Hunger Games* and other dystopian novels serves as a good example. "The Capitol" exploits thirteen districts, while the plans of President Snow remain secret for all participants in the Hunger Games. They are annually invited (forced) to compete in a lethal game where only one wins. Instead of cooperating, participants over-valorize their own skills and opportunities to become "the winner". Like in the labor market, where workers are moved to compete with other workers, *The Hunger Games* offers a false horizontality which is characterized by extreme competition. The Other's death implies, after all, that the participant has a better probability of continuing in the race. As I noted in my book, this is the main feature of Thana-Capitalism.

Lastly, the consumption of the Other's death cannot be limited to tourism; it can be replicated to almost all spheres of society and its institutions. The news is headed with the phrase, "these news may disturb the public's sensibility", containing not only scenes of violence and murders but the exhibition of corpses or the instant when the terrorist attack blasts the hotel. The audience is captivated by terrorism-related news because it affirms its aura of a privileged class. Though the images are appalling they cannot stop watching. This creates a strange interplay between terrorism and media coverage, because while the former needs mass publicity of their crimes, the latter gains further investors. This does not imply that citizens are sadists, but it is unfortunate that they are losing their empathy for the victims. Christopher Lasch as well as Richard Hofstadter envisaged this trend. When Lasch means "the culture of Narcissism" he signals the frustrations of an entire generation (with its care-takers) which developed a distorted sense of reality. Narcissists only move by internal dispositions, without any interaction with the Other. The Other's pain serves to maximize the pleasure in the narcissist character. So, Tzanelli's attempts to think of dark tourism as a form of empathy should be at least reconsidered. This point was widely validated by French writer Yves Michaud in his seminal book *Le Nouveau Luxe* (2013). He holds the thesis that contemporary society adopted a new type of luxury which cannot be shared or communicated. By the emulation of solipsist experiences, such as a safari in Africa or one day at the cheapest hotel, the new tourists are egoists, underpinned in their own narcissism while the sense of luxury supports their exemplary status. For this reason, as I have already clarified, the theory of Thana-Capitalism is neither a projection nor an extrapolation which can be universally applied. It describes an emerging trend which has come to stay. For example, *Investigation Discovery* (*ID*), one of the most prestigious documentaries watched daily by American viewers, narrates the biographies and stories of serial killers and how their victims are kidnapped, raped, tortured and finally killed. *Big Brother*, as it happens, shows a similar setting. A bunch of participants must live together in a house, making public all their miseries but covering their real interests. Each week, one participant is

forced to abandon the house, while the rest not only hope to be "the elected winner", but still conspire to defeat other participants. What is equally important, the
authority and hegemony of *Big Brother* is never questioned because the premise
of the "winner takes all" works. Another metaphor that explains how this society
functions is associated with what I dubbed the "syndrome of survivor", which can
be found in disaster movies. In the plot of much apocalyptic cinema, the disaster
comes because the society should be disciplined (by the Gods) for its greed or
any other sin. However, survivors feel they are special in view of the fact that the
Gods protected them, giving them a second opportunity to continue life on earth.
Although it represents part of the psychological resilience, which serves to protect
survivors from the trauma – likely this is the empathy to which Tzanelli refers –
unless strictly regulated, it may engender pathological forms of chauvinism and
nationalism. The survivor feels that after all the destruction, he or she is special,
outstanding or holding a sacred inheritance. This pathological character paves
the ways for the emergence of racist expressions where the alterity is subordinated. Narcissism, which is stimulated by the disaster films, not only disorganizes
but also undermines social ties. Henceforth as the theory of Thana-Capitalism
explains, death is figured in ambiguous terms. On one hand, the process of secularization instills some manifest fears of death, because it represents the end of
our entire material world, but at the same time, it cultivates a propensity to extend
life. On the other hand, figuring out life as a long race where only a few arrive at
the end, citizens feel pleasure for the Other's death because – in this way – they
are still in the race.

Tzanelli's review, most probably, exaggerates her trust in the role of dark tourism as a platform to interpret the Other's suffering. However, her review also
denotes some points which remain unresolved in my argumentation. In this book
I will resolve them one by one in the best way I can.

Conclusion

I have placed many examples that will help readers understand the concept of
Thana-Capitalism. Undoubtedly, the founding event that marked the start of a
new stage was 9/11. This event was widely covered, inaugurating a new tradition
which was historically enrooted in disaster cinema. Although the attack took the
lives of almost 3,000 innocent people, the smoke, destruction and the bodies falling from the World Trade Center shocked the US and the world. In this respect, the
photograph "The Falling Man" was taken by Richard Drew, a photographer linked
to Associated Press. However, almost 200 people fell to their death. This photograph was illustrative of the years to come as well as the need to gaze in detail
at how victims are systematically killed or blown up. While the terrorist cells
in former decades targeted important persons, now terrorism prefers to focus on
mobile citizens such as journalists, travelers or tourists. The 9/11 attacks may well
be interpreted from two different angles. On one hand, the attack was achieved by
employing the classic means of transport, which were the essential instruments of
mobility. To put this in other terms, airplanes were weaponized against the most

important icons of capitalism and the US. The World Trade Center symbolized the core of business and finances while the Pentagon was the place of military strategists. A fourth attack under way against Washington City was frustrated once Flight 93 was forcibly landed in Pennsylvania after passengers overpowered the hijackers. The original goals of Al-Qaeda were humiliation and the need to force the rest of the world to watch how the US was symbolically defeated. Of course, this attack caused not only a set of military invasions of the Middle East but also the orchestration of surveillance protocols in the name of security. After this event, life in the US substantially changed to the extent that some analysts manifested their concerns on the decline of democratic institutions.

Paradoxically, on the other hand, 9/11 woke up "a morbid character", which was already enrooted in American life. Over the years, popular culture in the US divided the world in two, we (the good democratic boys), and they (the undemocratic evildoers). This sentiment accompanied not only the liberal state from its outset, as we have already discussed in earlier sections, but also the social imaginary in America. It is safe to say that popular culture was prone to commoditize disasters as a form of entertainment. This exhibited a fertile ground for terrorism and particularly for Al-Qaida's courses of action. Basically, this created a curious interplay between the media and terrorism. While the former looked to attract further investors to improve their profits, the latter needed free advertising to publicize their crimes. It is possible that 9/11 did not create but accelerated the rise of a new stage of capitalism, where the Other's death is the main criterion of pleasure for the public. Thana-Capitalism, as I replied to Tzanelli, seems not to be an exaggeration but an ever-increasing trend which is neglected by interviewees. The idea that tourists – like contemporary citizens – are in search of novelty seems to be adopted by some scholars who placed too much attention on what interviewees said. The first dark-tourism research focused on the interviews and questionnaires applied to tourists just after the site is visited. Most probably, they replied they were there to learn more about the catastrophe or were simply looking for answers to their own lives. Hence they concluded that in essence dark tourism represented an anthropological attempt to domesticate death, in a hyper-secularized society. From their viewpoint the proof was clear, and what tourists said was a valid source of information. In view of this, since dark tourism played a leading role in the process of memorizing, communities constructed a dark heritage revolving around these disasters. Although a great portion of studies went in this direction, the thesis that outlines dark tourism as a part of heritage rests on shaky foundations. First and foremost, ethnography and anthropology show amply that sometimes interviewees do not tell the truth, because they need to protect their interests or because they are not cognizant of their emotions. Some interviewees may respond with energetic conviction while fundamentally there is no scientific causality with the studied object. In fact, what tourists think is not a validated hypothesis but only what they think.

Let me share one of my own experiences of the Republica de Cromañon, a nightclub in Buenos Aires devastated by a fire where 194 young people were killed. My ethnography took almost four years and I picked up a lot of false information that helped me to understand how the mind works and reacts to the disaster. One day, a teenager came close to me to talk about Cromañón. I originally started an informal interview with routine questions. He told me that he was homeless and that he had witnessed the tragic event The interview was more than I expected, in-depth and rich in details. He narrated and expanded my understanding of this issue, but once I compared the information with other key informants I realized he was lying. He never lived on the streets, nor was he a direct witness of the event. Quite aside from this, what he told me helped me during my ethnography to correlate specific indicators which were empirically validated. The same applies for the recent understanding my colleagues hold respecting dark tourism and their manifest reluctance to accept my theory of Thana-Capitalism. The main limitation was not given by what my key informant uttered, but the method I used. I was hearing without listening. Cromañón's lessons led me not only to think that the specialized literature on dark tourism was wrong, but to understand the obsession for gazing at death is not limited to the tourism industry, and it can be replicated in many other forms of cultural entertainment. *The Rise of Thana-Capitalism and Tourism* departed from this premise, and though there is still further discussion and questions around my argument, it is no less true that terrorism evinces this double hermeneutic circle that ushers citizens towards double standards on democracy.

References

Altheide, D. (2017). *Terrorism and the Politics of Fear.* New York, Rowman & Littlefield.

Ariès, P. (1975). *Western Attitudes toward Death: From the Middle Ages to the Present* (Vol. 3). Baltimore, Johns Hopkins University Press.

Bianchi, R. (2006). Tourism and the globalisation of fear: Analysing the politics of risk and (in) security in global travel. *Tourism and Hospitality Research, 7*(1), 64–74.

Bianchi, R., & Stephenson, M. (2014). *Tourism and Citizenship: Rights, Freedoms and Responsibilities in the Global Order* (Vol. 11). Abingdon, Routledge.

Biran, A., Liu, W., Li, G., & Eichhorn, V. (2014). Consuming post-disaster destinations: The case of Sichuan, China. *Annals of Tourism Research, 47*, 1–17.

Cohen, E. H. (2011). Educational dark tourism at an in populo site: The Holocaust museum in Jerusalem. *Annals of Tourism Research, 38*(1), 193–209.

Comaroff, J. L., & Comaroff, J. (2009). *Ethnicity, Inc.* Chicago, University of Chicago Press.

Debord, G. (2012). *Society of the Spectacle.* Canberra, Bread and Circuses Publishing.

Elias, N., & Dunning, E. (1986). *Quest for Excitement. Sport and Leisure in the Civilizing Process.* Oxford, Basil Blackwell.

Enders, W., & Sandler, T. (2011). *The Political Economy of Terrorism.* Cambridge, Cambridge University Press.

Germann Molz, J. (2012). *Travel Connections: Tourism, Technology, and Togetherness in a Mobile World.* Abingdon, Routledge.

Howie, L. (2011). *Terror on the Screen: Witnesses and the Re-animation of 9/11 as Image-Event, Popular Culture.* Washington, DC, New Academia Publishing, LLC.

Howie, L. (2012). *Witnesses to Terror: Understanding the Meanings and Consequences of Terrorism.* New York, Springer Nature.

Johnston, T. (2013). Mark Twain and the innocents abroad: Illuminating the tourist gaze on death. *International Journal of Culture and Hospitality Research, 7*(3), 199–213.

Kaelber, L. (2007). A memorial as virtual traumascape: Darkest tourism in 3D and cyberspace to the gas chambers of Auschwitz. *Ertr, e Review of Tourism Research, 5*(2), 24–33.

Keane, S. (2006). *Disaster Movies: The Cinema of Catastrophe* (Vol. 6). Wallflower Press.

Korstanje, M. (2016). *The Rise of Thana-Capitalism and Tourism.* Abingdon, Routledge.

Korstanje, M. E. (2018). *Terrorism, Tourism and the End of Hospitality in the West.* New York, Springer Nature.

Korstanje, M. E., & George, B. (2017). Death and Culture: Is Thana-Tourism Symptomatic. In *Virtual Traumascapes and Exploring the Roots of Dark Tourism*, Korstanje, M., & George, B. (eds). Hershey, IGI Global, 150–174.

Lacy, M. J. (2001). Cinema and ecopolitics: Existence in the Jurassic Park. *Millennium-Journal of International Studies, 30*(3), 635–645.

Lasch, C. (1991). *The Culture of Narcissism: American Life in an Age of Diminishing Expectations.* New York, WW Norton & Company.

Lash, S., & Urry, J. (1992). *Economies of Signs and Space* (Vol. 26). London, Sage.

MacCannell, D. (1976). *The Tourist: A New Theory of the Leisure Class.* Berkeley, University of California Press.

MacCannell, D. (1992). *Empty Meeting Grounds: The Tourist Papers.* London, Psychology Press.

MacCannell, D. (2001). Tourist agency. *Tourist Studies, 1*(1), 23–37.

Mair, J., Ritchie, B. W., & Walters, G. (2016). Towards a research agenda for post-disaster and post-crisis recovery strategies for tourist destinations: A narrative review. *Current Issues in Tourism, 19*(1), 1–26.

Mead, G. H. (1934). *Mind, Self and Society* (Vol. 111). Chicago, University of Chicago Press.

Michaud, Y. (2013). *Le nouveau luxe: Expériences, arrogance, authenticité (The New Luxury: Experience, Arrogance and Authenticity).* Paris, Stock.

Pietz, W. (1985). The problem of the fetish, I. *RES: Anthropology and Aesthetics, 9*(1), 5–17.

Pizam, A., & Smith, G. (2000). Tourism and terrorism: A quantitative analysis of major terrorist acts and their impact on tourism destinations. *Tourism Economics, 6*(2), 123–138.

Poria, Y. (2007). Establishing cooperation between Israel and Poland to save Auschwitz concentration camp: Globalising the responsibility for the massacre. *International Journal of Tourism Policy, 1*(1), 45–57.

Raine, R. (2013). A dark tourism spectrum. *International Journal of Culture, Tourism and Hospitality Research, 7*(3), 242–256.

Sather-Wagstaff, J. (2016). *Heritage That Hurts: Tourists in the Memoryscapes of September 11.* Abingdon, Routledge.

Séraphin, H., Butcher, J., & Korstanje, M. (2017). Challenging the negative images of Haiti at a pre-visit stage using visual online learning materials. *Journal of Policy Research in Tourism, Leisure and Events, 9*(2), 169–181.

Shondell Miller, D. (2008). Disaster tourism and disaster landscape attractions after Hurricane Katrina: An auto-ethnographic journey. *International Journal of Culture, Tourism and Hospitality Research, 2*(2), 115–131.

Skoll, G. R. (2007). Meanings of terrorism. *International Journal for the Semiotics of Law-Revue internationale de Sémiotique juridique, 20*(2), 107–127.

Sönmez, S. F. (1998). Tourism, terrorism, and political instability. *Annals of Tourism Research, 25*(2), 416–456.

Sönmez, S. F., & Graefe, A. R. (1998). Influence of terrorism risk on foreign tourism decisions. *Annals of Tourism Research, 25*(1), 112–144.

Stone, P. R. (2012). Dark tourism and significant other death: Towards a model of mortality mediation. *Annals of Tourism Research, 39*(3), 1565–1587.

Stone, P., & Sharpley, R. (2008). Consuming dark tourism: A thanatological perspective. *Annals of Tourism Research, 35*(2), 574–595.

Tzanelli, R. (2016). *Thanatourism and Cinematic Representations of Risk: Screening the End of Tourism* (Vol. 176). Abingdon, Routledge.

Tzanelli, R. (2017). Bios and Thanatos in leisure: A critical review of Maximiliano E. Korstanje's The Rise of Thana-Capitalism and Tourism. *European Journal of Tourism Research, 17*, 269–271.

Urry, J. (2002). The global complexities of September 11th. *Theory, Culture & Society, 19*(4), 57–69.

Urry, J. (2004). The "system" of automobility. *Theory, Culture & Society, 21*(4–5), 25–39.

Urry, J., & Larsen, J. (2011). *The Tourist Gaze 3.0*. London, Sage.

Walters, G., & Mair, J. (2012). The effectiveness of post-disaster recovery marketing messages—The case of the 2009 Australian bushfires. *Journal of Travel & Tourism Marketing, 29*(1), 87–103.

White, L., & Frew, E. (2013). *Dark Tourism: Place and Identity: Managing and Interpreting Dark Places*. London, Routledge.

4   Is torture enough?

Introduction

In the first chapter, we adamantly discussed the role of torture in democracy. In so doing, we placed in dialogue the works of Daniel Feierstein and Jaime Malamud-Goti. Although both approaches sounded very interesting and pertinent for the present book, their outcomes are historically determined by the 1970s. In fact, part of the human rights violations, which were documented and debated by Feierstein and Malamud-Goti, were perpetrated by military forces during dictatorships. The fertility of this discussion posits that democracies appeared to be immune to the advance of terrorism. American political scientists of all stripes outlined that terrorism resulted from weak democracies or authoritarian governments (Butler 1976; Alexander 1976). Time has proved this was not the case, at least for the US, with the allegations of victims of torture after the application of the Patriot Act (Skoll 2016). What would happen if democracies were not immune to terrorism?

Today's terrorism has substantially shifted and the role of military forces and the repressive mechanisms of state are accompanied by more intrusive surveillance technologies (Lyon & Bauman 2013). After 9/11, the US not only sanctioned the Patriot Act, which modified the existent juridical background, but also encouraged the private sector in the construction of Supermax (maximum security) prisons, where the inmates' rights are systematically violated. However, despite the resources the US and other main powers have, the tactics to struggle against terrorism seem to be less efficient than in other periods. This leaves a philosophical dilemma regarding the efficiency of torture in the contemporary world, which will be addressed in this chapter. It is vital to mention that torture appeared to be not enough – if not counterproductive – to reduce the virulence of terror. The Patriot Act reminds us that the most horrendous crimes might be legally sanctioned with some consensus. This poses the question: to what extent does the majority have the right or the correct answer (Skoll 2016)?

Against this backdrop, the leaders of ISIS, Abu Musab al-Zarqawi and Abu Bakr al-Baghdadi, were subject to torture and countless humiliations in Abu Ghraib prison. The different punishments and human rights violations borne daily by inmates in these prisons reacted energetically in a sentiment of hostility against the US that allowed the rise and expansion of one of the major concerns of modern politics: Islamic State. This chapter centers on the need to understand the nature of terrorism and the inefficacy of torture to reveal information about the next attack. Furthermore, it places the theory of "lesser evil", which was originally formulated by the liberal writer Michel Ignatieff, under a critical lens.

Somehow, as A. K. Cronin observes, modern terrorism contours its power from the legitimacy of the nation-state. With the passing of time, the attitude of the public regarding terrorism have changed. While in the 1970s terrorists were considered as guerrillas who resisted the nation-state, nowadays, they are considered evildoers, demons or psychopaths who place the public in jeopardy. To wit, Cronin acknowledges that attitudes have changed according to some developments. At first glimpse, the public is more sensitive to news and exposed to how terrorists kill civilians. Secondly, the media technologies package and disseminate terrorism-containing news in minutes worldwide. Over the years, it is no less true that some radical cells were trained and sponsored by other states. For example, the US trained Osama Bin Laden and Al-Qaeda as long ago as the Soviet invasion. In the terrorist's mind, success consists in the greater access to more "lethal means" to cause panic in the major proportion of society (Cronin 2011). Rather, Luke Howie (2012) has stressed that terrorists do not want a lot of people to die, they want a lot of people watching. In fact, terrorists would emulate the role of celebrities in popular culture, captivating the audience's attention to fulfill their goals. One of the main problems in understanding the issue relates to the lack of knowledge regarding how terrorists think. Since terrorism is cataloged as a serious crime, any contact or key information about the biography of a suspect of terrorism should be shared with the State. This discourages many ethnographers and ethnologists from entering these types of fields. As a result of this, the TV screen is fraught with pseudo-experts who nourish the discourse of the status quo, which calls for more intervention in the so-called "failed states" or tougher penalties on terrorists. Paradoxically, the media culture intensifies the efforts by covering terrorist acts, opening the doors for the creation of a climate of paranoia as never before.

In consonance with this, Lisa Stampnitzky (2013) remembers that terrorism has received different treatment and connotation over time. It ranged from an "illegal violence towards an evil act". As she puts it, before 1972, the problem of political violence was strongly associated with insurgency, which derived (following analysts) from different socio-economic frustrations and cleavages. No less true was that in Europe many of the political experts dedicated to studying counter-insurgency were former members of military forces, while in the US academicians occupied such a position. It is important not to lose sight of the fact that those experts, who were originally advocated to understand the roots of the insurgency, dangled the possibility that violence derives from grievance, or psychological frustrations resulting from uneven wealth distribution. The question is whether poverty in the developing world played a crucial role in delineating the borders of political discourse with respect to the preventive policies the State should adopt towards "guerrillas". However, from the 1970s on, the position was radically altered towards a much more complex situation, where terrorists were framed as "evildoers", insane or mentally ill. This happens for two main reasons. Firstly, the multiplication of attacks against defenseless people, but secondly and most important, Americans witnessed how the rise of a new sentiment of hostility fed a radical discourse against them. To put this bluntly, the United States and its citizens were situated as the main targets of international terrorism worldwide. At the time that experts advanced their conclusions, the invention of new hypotheses and outcomes, which led to the maturation of the discipline, the meaning of terrorism was radically altered. Terrorism was conceived as an "irrational activity" or as a major threat which should not be studied but eradicated. As a result of this, some old neo-pragmatist narratives were installed in public opinion, demonizing terrorists as the enemies of states. This voluntarily or not ushered Americans into a quandary, which was to decide between a process of securitization and the checks and balances of power. This point was brilliantly explored by other scholars including David Altheide (2006), Luke Howie (2012), Johnathan Simon (2007), Mahmoud Eid (2014), Noam Chomsky (2015) and Geoffrey Skoll (2016) among others. All these voices reach consensus in indicating that governments appealed to internal or external dangers – like terrorism or local crime – to overcome the juridical obstacles mined by Republicanism. To some extent, the executive branch manipulated the discourse against terrorism to promote economic policies that otherwise would be rejected by the electorate. For these scholars, terrorism – far from being a threat – would be an ideological discourse orchestrated to subdue the workforce. However, other academicians like Michael Ignatieff confirm the polemic thesis that torture, even if it is regulated by the law and constitution, may be a useful instrument to defeat terrorism. In the following sections, we shall deal with Ignatieff's theory as well as philosophical discussions revolving around torture, democracy and terrorism.

The doctrine of lesser evil

Michael Ignatieff is a well-known Canadian political scientist, who has studied issues related to democracy, politics, human rights policies and terrorism over recent decades. Though he is educated in the liberal arts, Ignatieff has been influenced by many academic waves and theorists. At a closer look, he combined an interesting career in politics with academia. In this section, we shall explore of some his most significant works to learn more about the subtle connection between liberal thinking, which is aimed at protecting the rights of individuals, and torture. Ignatieff's metamorphosis gives further understanding of how terrorism and the discourse around "the war on terror" affect the life of lay-citizens, subordinating public opinion to more radical (populist) claims.

At first glance, in *Human Rights* Ignatieff (2001) clarifies that human rights are not good or bad, but necessary. The end of WWII envisaged the necessity of legislating against crimes against vulnerable citizens and minority groups. Nazi Germany devoted the available material resources to exterminate what they dubbed as "unter-menschen" (sub-humans). Nonetheless, the extreme violence exerted against these minorities was possible because they were dispossessed from their inherent legal rights. Henceforth nations looked for rapid and consented legal answers to these atrocities. The successive covenants and international agreements celebrated among nations were oriented to prevent the crimes against civilians and vulnerable groups by reasons of religious affiliation, nationality or ethnicity. After all, the sanction of human rights not only recovered the identity of the self, which was deprived by the Nazis, but it aggravated the situation for many ethnicities, which during the twentieth century were assassinated by their nation-states. Although nation-states signed the international agreements at least, no efforts were made to ensure the human rights of ethnic minorities within the borders of nationhood. This raises some important questions: what are human rights and how do they function? Ignatieff asks.

In view of the juridical chasm, many NGOs and militants began to promote and expand the coverage of vulnerable groups throughout the globe. Asking for assistance from business organizations such as the World Bank, they tried to impose an agenda, which obscured more than it clarified. The situation of thousands of asylum seekers, migrants and exiles were never improved but worsened. Ignatieff recognizes that the American exception paved the way for a much deeper resentment against the US because of its lack of support in contexts of extreme violence, ethnic cleansing, genocide or slaughters, but – as he describes – the human rights activists cannot be blamed for the situation they involuntarily created. There is an idolatry of human rights that ignores the real nature of the phenomenon. In this respect, human rights should be framed on the table of international relations to reach multilateral consensus. Only deliberative democracy may resolve the disputes that come from politics. As one state forces others to respect human rights, the right of sovereignty declines, as Americans amply accept. Here some questions arise: do the needs of security outweigh democracy? How can democracy deal with terrorism? Is torture a solution?

The second book to inspect is entitled *The Lesser Evil: Political Ethics in an Age of Terror*. It is safe to say that this project is probably one of the most intriguing and polemic interrogations by Ignatieff regarding the role of democracy in the War on Terror. From the benefits of hindsight, Ignatieff writes that democracies, in contexts of disorder and chaos, naturally reduce their protection of individual rights. While the constitution, which is considered the cornerstone of democracies, has no application in contexts of danger, when the conditions of emergency increase without constraints there is a serious probability that dictatorships will emerge. As Ignatieff observed, terrorism favors a climate of political instability, which remains a fertile ground for totalitarian governments. To avoid this, democratic government should operate within the borders and mandates of the law. But what happens when a suspected person has vital information about the next terrorist attack? In this case, some analysts suppose that torture not only saves lives but also is vital to efface the scourge of terrorism. Unless the torture is regulated by law, the government runs the risk of falling into serious human rights violations, Ignatieff claims. The institutions of checks and balances and the control over security forces ensure prevention of future abuses or violations of individual liberties. In this conjuncture, States should retain the capacity to intervene not only in national but in territorial dimensions. Sometimes, States make an intervention in a national sphere, suspending constitutional liberties, while others employ a selective monitoring, which means the suspension of liberties of some ethnic minorities. Lastly, the lesser evil doctrine probes that democracy is efficient in the legal regulation of the action of militaries as well as other security forces to grant the security of all citizens (Ignatieff 2013).

In 2005, Ignatieff edited a new work, which invited academicians and scholars to discuss the American exceptionalism. In the preface, he argues that since the end of WWII America has conducted an exemplary role in the human rights field. However, paradoxically, the American Government – no matter if Republican or Democrat – refused the intromissions of other States in domestic issues. As he cites, "This combination of leadership and resistance is what defines American human rights behaviour as exceptional, and it is this complex and ambivalent pattern the book seeks to explain" (Ignatieff 2005: 1).

What seems to be exceptional is not the reluctance to be blamed by other nations for violating human rights in other nations, but how America combines a sentiment of guilt that is crystallized in a leading role in the promotion of human rights elsewhere beyond the borders of the US. Ignatieff explains that the sense of exceptionalism played a historical role placing Americans outside the judgment of any other nation. Under the auspices of the doctrine of self-determination, from its inception, America understands any direct or indirect constraint to its sovereignty is inadmissible. Of course, this allows some abuses that violate the protocols of the Geneva Convention, for example in Guantanamo and Abu Ghraib, or simply the omission of the Kyoto Protocol to reduce greenhouse gases in the atmosphere. Ignatieff resolves not to deal with the history of Puritanism or religion to explain these phenomena because exceptionalism and unilateralism go along different paths. According to his view, American exceptionalism is based on three constituent elements which deserve to be mentioned here.

> First, the United States signs on to international human rights and humanitarian law convention and treaties and then exempts itself from the provision of explicit reservation, non-ratification, or noncompliance. Second, the United States maintains double standards: judging itself and its friends by more permissive criteria than it does its enemies. Third, the United States denies jurisdiction to human right law within its domestic law.
>
> (Ignatieff 2005: 3)

While Americans often take the lead in signing international agreements, in doing so they fail systematically to abide by their requirements. Centered on a double standard, the US negotiates treaties trying to create a hegemonic network, in which case, as Ignatieff notes, it affects its credibility in the struggle against terrorism or other major threats. Ignatieff resolves this point, mentioning that the US constitution appears to be one of the oldest in the world, and US rights guarantees have been used in the past as a critical instrument against the government. Political parties are more consistent than other nations in controlling the executive branch because American democracy serves as the natural evolution of checks and balances. In that way, Ignatieff presents the problem and at the same time, the solution. Though the US sometimes vulnerates the autonomy of some nations, even practicing torture in Supermax prisons, a would-be active public opinion claims the government is behaving ethically.

Another perspective

Ignatieff received a lot of critical comments, many of them related to his exaggerated trust in American democracy as the best of feasible worlds, and his endorsement of what he calls "a legal torture" (Souter 2009; Morefield 2008; Cistelecan 2012; Korstanje 2013). It is tempting to say Ignatieff works his theory from a distorted image of democracy, which does not really adjust to historical facts. To set an example, French Philosopher Robert Castel (1997) presented interesting research validating the assumption that the Industrial Revolution undermined social ties and traditions, which were originally coined in the Middle Ages. The introduction of liberty was conducive to the liberation of workers, which was rechanneled towards a systematic exploitation of women, men and children in the factories. Castel successfully describes how the triumph of capitalism derived from the adoption of freedom as a universal concept, but creating a gap between the lower classes and the ruling elite. The economic changes brought by capitalism were adjoined to cultural changes. Free from the traditional liaison with their masters, the peasants were pressed to sell their labor. Moved exclusively by psychological needs, peasants migrated to the urban cities where they became impoverished. Poverty and the difficult living conditions of thousands of citizens in the urban area was a result of the introduction of liberty as the main ideological core of capitalism. By this token, Cornelius Castoriadis (2006) has shown that democracy, far from being the form of government contemporary societies today valorize, took a different shape in Ancient Greece. Neither the authority of the King nor the Senate was subject to elections; democracy, rather, was based on the administration of "demos" resources through which any free citizen can call the assembly to derogate a law if it is unjust. As Korstanje (2018) puts it, the modern conception of liberty is given by the liberalization of workforces. The Anglo-Saxon empires (the United Kingdom and the United States) not only equivocated about the meaning of democracy but impeded the lay-people from active participation in the sanction of laws. The old liberty of the Greeks set the pace to a new one, corporative and manipulated by the elite to stimulate mass consumption. Therefore, we must speak of Anglo-democracy as the political expression of republican values as well as the laws which are sanctioned by corporations. Thirdly, Ignatieff's diagnosis on human rights leads inevitably to the dictatorship of a supra-state. Ignatieff glosses over the fact that the paradoxes of human rights are not given by the militants as he precludes but by the design of the same nation-state. From its onset, the liberal state reserved the use of two major capacities: self-protection against the attacks of a third state and the protection of its citizens. Far from any negotiation, in which case Ignatieff assumes democratic states are more sensitive to negotiate than totalitarian ones, States will always do their best to preserve the right of self-determination with respect to the intervention of a third state. So we must ask why the self-determination of the democratic nation is superior in legal terms to authoritarian ones.

Once again, the political exemption is not given by a sedimentation of cultural values, as Ignatieff insists, but as a vicious circle orchestrated by the main powers to keep a hegemony over the periphery. This was validated by Michael Freeman (2011) in his caustic critique of human rights as simple inventions which are evoked in context of an emergency. What the theorists of human rights failed to resolve seems to be why some rights are more important than others. In consonance with this, J. Albrecht-Meylahn (2013) criticizes Ignatieff when he writes that the use of violence is proportional to pervasive counter-effects. Based on the contribution of S. Zizek, any revolution is condemned for the same cruel acts previously criticized by them before the coup. This happens because of two primary motives. Firstly and most important, physical violence may fluctuate, but if the systemic violence is not extirpated, the material asymmetries that triggered the revolution persist. On this logic, Ignatieff is unable to resolve the paradoxical situation of torture which the most democratic nation of the world, the US, today encourages.

Secondly, Gutmann (2001) tackles the text of Ignatieff alluding that his problem is the concept of sovereignty. It is based on the need to believe in negative liberty. Ignatieff's account is troublesome for the following reasons:

- If in the Middle East people culturally are pressed to decide to stone a woman because of infidelity, why does the US condemn such punishment?
- Gutmann is correct in confirming that the discourse of sovereignty is often invoked by those criminals who violate human rights.
- As formulated by the West, human rights looks to develop an ethnocentric view of self-determination, a chauvinistic expression of nationhood that leads to the theory of "exemption".

In a nutshell, Gutmann criticizes Ignatieff because of his obsession to believe that self-defense and self-determination are natural values *per se*. If States are drawn to grant the well-being of their members (the protection of the weak), human rights confer legitimacy to the law of the stronger. Gutmann considers that nationalism is the main problem of human rights theory. The way Ignatieff understands the legal jurisprudence poses a serious challenge for liberal thought and, of course, this awakened a serious criticism. Lastly, A. Appiah (2001) contemplates that rights are not subject or individual but applied collectively. Technically, rights are not derived from deliberation, nor individualism, as Ignatieff says, but from the collective spirit of community.

Unveiling the Lucifer effects

Philipp Zimbardo is a well-renowned psychologist, who served as the principal scientist of the Stanford Prison Experiment, which was a simulation of life in prison. Seventy-five students accepted Zimbardo's invitation to take part in an experiment oriented to study perceived power. The group was randomly divided into two bands, officers and inmates. The outcome was that many of prisoners accepted passively the maltreatment and abuses of officers. The participants internalized their roles until the experiment was discontinued after Christina Maslach, a graduate student in psychology objected to the morality of the experiment. Quite aside from the ethics and criticism of this polemic text, Zimbardo's findings are applicable to studies to understand the relation between inmates and guards. In 2007 and pressed by the news of torture in Abu Ghraib, he published a book, *The Lucifer Effects*, which recapitulates his vast experiences in the Stanford Prison Experiment (SPE) in the new context. As this name indicates, he is obsessed with knowing why and how a good person turns evil. While lay-people are prone to enhance their ego, feeling they are special or outstanding, as Zimbardo contends, the rules limit their behavior. The classic definition of evil or evildoing should be reconsidered via the circles of power, how rules are formed and accepted or negotiated by the in-group members. Any behavior shifts automatically when the self is introduced to the new norms and authorities. Zimbardo interrogates himself on the nature of evil, asking for reconsideration of its instrumental essence. In fact, he clarifies, we are accustomed to thinking of evil as any act which is directed to harm, abuse or dehumanize the "Other". Fundamentally, such a definition overlooks the fact that evil represents an incremental thing explained by the means-and-end logic. Evildoers, in this vein, are not monsters or supernatural entities, rather they are humans, who not only are subject to rules and contexts but also do not want to be excluded from the group. This disposition attitude is certainly given because they need to be accepted by their peers. This point spans from SPE to Abu Ghraib, suggesting that a good person becomes an evildoer when the rules are substantially changed. He offers as a valid example, the history of inquisition. While the Catholic Church promoted inquisition, it was not seen as an act of torture and dehumanization; it was an act of God. Starting from the premise that power is based on the convergence of ideology and the self, Zimbardo acknowledges that the legitimacy of the ruling elite consists in its efficacy of manipulating and selectively socializing the citizens into a system of values, which seems to be previously drawn. This happens because when the self is trapped between fear and doubt, it follows the wishes of the majority. Hence society needs an alter ego, a staunch enemy which is stereotyped to instill fear. The alterity signals to the frightened "Other", who may jeopardize the well-being of the collective soul.

The most extreme instance of this hostile imagination at work is of course when it leads to genocide, the plan of one people to eliminate from existence all those who are conceptualized as their enemy. We are aware of some of the ways in which Hitler's propaganda machine transformed Jewish neighbors, co-workers, even friends into despised enemies of the State who deserved the "final solution" (Zimbardo 2007: 11).

The atrocities, slaughters and genocides against innocent civilians may very well be committed by ordinary people if the social conditions are given. The SPE amply shows that part of the abuses by guards is legitimated in their impunity and the state of conformity, which are engulfed in the existent ideology. Good people in bad climates not only adapt their cultural and ethical values but also appeal to a process of dehumanization (over the victim) to justify their acts. To a major or minor degree, Zimbardo's work sheds light on the psychological understanding of role expectation and the dialectics between dominators and dominated. The sentiment of subordination from the latter respecting the former depends on the efficacy of ideology, which facilitates the involved participants to internalize their roles. One of the paradoxes of Lucifer effects seems to be that disobedience corresponds with the reasons and the reaction to the imposition of law. Since guards need to keep order, the immediate reaction of the inmates is to generate chaos and rebellion. What Zimbardo reminds us is that any order denotes a symbolic essence which forms what psychologists know as personhood. When the "Other" is dispossessed of its rights and the condition of "being human", its ontological security will be at stake.

> People usually select the settings they will enter or avoid and can change the setting by their presence and their actions, influence others in that social sphere, and transform the environment in myriad ways. More often than not, we are active agents capable of influencing the course of events that our lives take and also of shaping our destinies. Moreover, human behavior and human societies are greatly affected by fundamental biological mechanisms as well as cultural values and practices.
>
> (Zimbardo 2007: 320)

The degree of violence in society should not be explained in moral or philosophical terms, Zimbardo ends. Otherwise, we are doomed to repeat the history of genocides that the twentieth century witnessed. History evinces that the meaning of acts (even paraphrasing Arendt's banality of evil) works according to the local system of values. Institutional goals, and not morality, rule the behavior of lay-citizens. Then if we accept torture as a mainstream cultural value (to fight against evildoers), any democratic order may commit horrendous crimes.

Terrorism and the nature of fear

In the novel *The Climate of Fear*, Wole Soyinka (2005) argues that 9/11 was nothing new. Many other citizens from the periphery faced similar or worse violence decades before this event. However, no less true was that from that moment onwards, public opinion was involved in a spiral of panic as never before. Soyinka believes the world has faced extreme situations of panic before 9/11 ranging from Nazism and the Second World War to nuclear weapon testing. One of the aspects of global power that facilitates this feeling of uncertainty seems to be the lack of a visible rivalry once the USSR collapsed. The politic terror promulgated by States diminishes the dignity of enemies. These practices are rooted in a territory but paved the way for a new form of terrorism which ended in the World Trade Center attacks. It is incorrect to see 9/11 as the beginning of a new fear but as the latest demonstration of the power of an empire over the rest of the world. Mass communications, though, transformed our ways of perceiving terrorism even if it did not alter the conditions that facilitate the new state of war. Soyinka's main thesis seems to be that the war on terror commoditizes and disseminates fear under two significant assumptions. On one hand, the audiences understand that the terrorist attack hits anytime and anywhere, undermining the credibility of the nation-state. On another, there are no geographical borders that can stop the attack nor any ethical limits respecting the vulnerability of non-combatants. In fact, one of the aspects that makes terrorism frightful is that the violence is directed against non-combatants or innocent civilians. The nature of fear is framed by greed or the need for domination, which tends to dehumanize the "Others" as a means for the achievement of their own goals (Soyinka 2005). Corey Robin (2004) pays heed to the role of fear as the architect of politics. Reviewing ancient and medieval sources, he assumes that the legitimacy of the State rests on two important assumptions. Firstly, an external enemy which is easily demonized in order to forge a social cohesion. Secondly, this outside danger serves to domesticate the conflict and discrepancies internally. Through a historical revision, Corey understands that modern nations work through a double-heuristics which combines fear and loyalty. To put this brutally, in the capitalist system, the need to disorganize worker unions was conducive to the interests of capital owners. The invention of an external foe leads rank-and-file workers to cede their claims or in some extreme cases, helps the government in dismantling the juridical representation of unions.

For many political scientists, Thomas Hobbes is considered one of the fathers of the modern state. This does not mean that Hobbes imagined the contemporary state, but his conception of power and the *Leviathan* illuminated modern thinkers in discussing the dichotomies of the modern nation-state. With England on the brink of civil war, Hobbes proposed a theory to understand not only the degree of surrounding violence but how people pass from the state of nature to civilization. In Hobbes's insight, the state of nature was fully marked by two contrasting tendencies. People desire what the other possesses while at the same time they are afraid others will take their own properties. To avoid the "War of All against All", Hobbes sustains, lay-persons endorse the monopoly of violence in a third entity, the Leviathan, which is no other than the State. The Leviathan, which is legally legitimized by the power of law, castigates those who violate the rules and the law in order to preserve the social peace (Hobbes 2006). Needless to say, his genius shed light on dark matter ethicists did not dare to resolve: the troubling relations between security and liberty. According to his stance, freedom was only a rumination, a fiction which cannot be grasped by humans. Most probably, his legacy rested in the fact that he pointed out that fear never disappears, as ethicists insist, but is mutated to other subtle forms or what is clearer, the fear remains encrypted into the ideological core of capitalism.

Despite its clarity and profundity, Hobbes was widely criticized by a great mind, who was worried by the relation of power, democracy and state: Michel Foucault. Unlike Hobbes, Foucault conceived power as a circular and exchangeable commodity, which is enrooted in a discursive narrative. To put this in other terms, he toys with the idea that power circulates everywhere, forming and shaping persons, institutions and discourses. He preliminarily addresses the connection between power and knowledge. The concept of truth, as it is imagined by the social imaginary, not only does not exist but is socially constructed by the ruling elite, which means those who historically took the power. Like history, the truth obeys the mandate of the hegemonic voices while others (peripheral) were repressed, silenced or pushed towards the borders (Foucault 1982). It is important that social scientists play the role of archeologists, who unveil the meanings of covered knowledge to confront and understand any established thought (Foucault 2012). Furthermore, in the seminal book *Society Must Be Defended*, he claims that the disciplinary mechanism is not strictly given by coaction. Rather it connotes to the economy of scarcity, where prices are fixed depending on the availability of commodities. The society appeals to the efficacy of discipline to inoculate external threats into "internal risks", as Foucault reminded us. He coins the metaphor of the vaccine for readers to comprehend how risk really works. The vaccine stands to the virus in the same way as risk stands to dangers. For Foucault, the vaccine should be technically understood as an inoculated virus, which is controlled once introduced into society (Foucault 2007). When Foucault uses the term controlled, this equals that the danger (virus) is dispossessed of all destructive features. Any external danger is disciplined and packaged in the form of risks in order for the economic order to be preserved. This concept will be vital to understanding the intersection of terrorism and the modern work (in the following chapters) (Foucault 2003).

One of the specialists in Foucauldian studies, Wendy Brown (2008), reads the complexity of his thought in a tripartite model, which is formed by "the sovereignty", the "commodity" and the "repressive". In sum, the sovereignty model signals to the power that is exercised beyond the persons and institutions. In sharp contrast to Hobbes and Rawls, Foucault strongly believed that the sovereignty is revealed as the bulwark of power, not as a consequence. This suggests the sovereignty is given as a fiction, which is internalized by the bodies, institutions and social relations in the same way. The commodity model rested on an economic understanding of power that is transferable or exchangeable among the individuals who form part of society. The commoditization of power, which applies in monarchies and democracies, derives from the ability of the ruling elite to create hegemony and the sense of commonness (sovereignty). Lastly, the repressive mode exhibits the prohibition or constraints over the subject. In opposition to liberals who defined liberty as the escape from oppression, Foucault says that the concept of liberty cannot exist outside the action of power and coaction. As a mere fiction, freedom is regulated by the disciplinary mechanism of the State.

As the previous argument shows, Noam Chomsky (2011) claimed that thinking of terrorism as the instrument of the weakest party not only is a mistake but also has echoes of biased diagnosis by the side of some intellectuals. The powerful groups have taken advantage from 9/11 because they monopolize the ideological apparatuses of the State. Basically, 9/11 was enthusiastically accepted as a founding event while other similar events like the massacres in Buenos Aires in 1992 and 1994 were certainly ignored.

Democracy and torture reconsidered

Most likely, there is no better example to describe the difficult situation regarding human rights in the US than the expansion of Supermax prisons, where thousands of terrorist suspects are systematically jailed without a fair trial. Doubtless, the attacks on the World Trade Center and the derived anxieties were fertile grounds for the acceptance of more repressive forms of control than in other decades. This is exactly what Derek Jeffreys (Professor of Humanistic Studies and Religion at the University of Wisconsin) thinks, who published in 2013 a very interesting work on the theme. With the benefits of hindsight, he focuses on the dilemmas of torture and the violence the inmates daily face in these types of control-units. The war on terror ignited a hot debate revolving around the tortures in Guantanamo Bay and other prisons. Though originally Obama promised to end illegal practices in Supermax prisons, it never happened. This awakened countless critiques against his administration and the role of army forces abroad. Although journalism played a vital role in the crusade against the executive branch, after further review, both sides agreed and endorsed the efficacy of Supermax prisons. Jeffreys reviews the literature with the strong focus on "solitary conferment", and the psychological impacts it has for people. In its history, as Jeffreys adds, the United States was characterized by developing a repressive system, which recreates a condition of extreme racialized violence within prisons. As a result of this, solitary confinement surged as a more "efficient" disciplinary instrument to keep control in the penal system. As Jeffreys puts it,

> Historically, American prisons often employed physical torture. For example, in the 1840s at the Sing Sing prison in New York State, inmate were subjected to water torture. Jailers released a solid stream of cold water from a great height, literally beating prisoners with water.
>
> (Jeffreys 2013: 35)

But things changed and now authorities use and abuse technologies for the same ends. In the Supermax, inmates are isolated for months or years, without any type of contact with other inmates or guards. They often suffer sleep deprivation and another sort of psychological torture that creates what has been described as the "golden cage problem", which causes serious psychological disorders. After the denunciations aired in 2004 of abuse and torture by the military forces in Abu Ghraib which included sexual abuse, humiliations, maltreatment, battery and aggravated assault among many other human rights violations, Supermax facilities have been disposed not to castigate all inmates, but rather, starting the protocols for a "selective isolation" of those who are extremely violent or show serious problems in living with others. As Jeffreys concludes, Supermax is today a disciplinary instrument to dehumanize the prisoners in the name of security and democracy. In this respect, Pilar Calveiro (2012), an Argentinian sociologist, alerts us that in the current stage of capitalism the world has changed. She toys with the belief that nations are suffering a radical reorganization, which is based on the neoliberal doctrine. Calveiro contends that a new violence – over the bodies of marginal minorities – is legitimated and administered by the States. While the war on terror leads global audiences to solicit harsher penalties in the Global North, local crime plays a similar function in the Global South. In both cases, the ruling elite expands the juridical object, which means the exact crimes which will be castigated, while the penalties become more severe. As stated, Calveiro traces her analysis back to the Cold War. Particularly, the Soviet Union was a counter-balance power for the hegemony of the US. In a bipolar world, both sides concentrated their power through technology and wealth. Once the Soviet Union suddenly declined, the US became the only super power on the planet. Calveiro reminds us that neoliberalism (as a discursive metaphor created to subordinate the periphery) from the 1990s on encouraged a fierce privatization of public space leaving open the intervention of banks and financial corporations in matters of state. The market and private interests arrived in politics, coopting politicians and exerting the monopoly of the State. This led to a difficult position because democracies were emptied of any content, and converted into pseudo-democracies. As a result of this, the business corporations enhanced their legitimacy using fear as a form of dissuasion. Citing Hardt & Negri (2001), she adheres to the idea that we live in a globalized empire, commanded by capital, where protests and dissidence are seriously criminalized. Abroad, States struggle against terrorism while internally worker unions and workers are coerced to accept the neoliberal policies. At the time terrorism is seen as a serious threat, few or scarce definitions really say what terrorism is. It paves the way for the rise of abstract and diffuse definitions which are oriented to be applied anytime and on any citizen. Calveiro's argument polemically theorizes about the need to understand terrorism from another angle. Her main thesis is that terrorism serves as a discursive narrative oriented to legitimate the exploitation of liberal states and their exploitative practices. What would be interesting to discuss here is the connection between terrorism and States. Paradoxically, the most democratic nation in the world, the US, conducts practices that violate international rights. The forced disappearance of prisoners who lack any rights to legitimate defense in a trial is accompanied by acts of torture that are silenced. To understand why the US is equated with totalitarianism, Calveiro said that Arendt was right when she confirmed that one of the aspects of these regimes was the production of non-persons. Totalitarianism introduces a hegemonic discourse, which nobody rejects, where the sense of humankind is annulled. Citizens embrace totalitarianism to fuse their loyalties to a great leader, while others (dissidents) are situated under the line of humanity. In view of this, the 9/11 attacks – as Calveiro avers – accelerate a long-dormant trend which was encapsulated in the roots of neoliberalism.

Though we find Calveiro's text very illustrative and interesting with respect to the compliance of states with corporations, as we have already evinced in earlier chapters, there is nothing like a direct intervention by the market in politics and matters of government. They are two sides of the same coin. From its inception, the liberal state was created to foster, protect and monitor the market. This means that the market and the nation-state are the two main pillars of capitalism. A capital owner would never amass their wealth without the laws sanctioned by the government to care for their interests, or as Karl Marx (1983) brilliantly said, the conception of commodity fetishism cannot be performed without the support of the law and State. The idea that the market avoids the law because it mimics (or infiltrates) the State resulted from an old Latin American romanticism, which was forged in the 1970s but without any rational basis.

Deliberative democracy

A great portion of studies focused on the role of "deliberative democracy" as a valid instrument useful in the fight against terrorism or dealing with cases of torture and abuse (Chappell 2012). However, one of the troubling aspects of the term consists in the fact that the majority never take part in a public debate or participate actively in a political party. It opens the doors to question, as Z. Chappell does, what is wrong with deliberative democracy?

The lack of empathy of citizens for politics depends upon their constant disillusionment in view of what their representatives do. Chappell (2012) argues convincingly that the deliberative democracy is determined by an adversarial logic, in which case the common well-being is tagged behind the interests of professional politicians.

By this token, Jurgen Habermas defines terrorism as a communicative glitch, which means a disruptive process impeding a correct communication among the involved parts (Borradori 2013). Following his previous assumption regarding the deliberative democracy, Habermas alludes to the creation of an international organism which monitors and places all sides in a genuine dialogue. Centered on an international cosmopolitanism, he eloquently suggests that deliberative democracy opens some forms of discussion which can be compared to the Kantian thesis that the world tends to a perpetual peace. The deliberation seems to be often governed by equal conditions of existence that lead to cooperation and mutual understanding. All parties have the right to put forward a theme for conversation as well as negotiate the rules on the discussion. Habermas is convinced "the deliberation" – to set an example among nations – is a prerequisite towards a "transnational democracy" (Habermas 2015; Dryzek 1999). However, as the philosophers who are detractors of deliberative democracy object, while globalization – as a cultural project – expanded across the globe, poverty not only persisted, it was notably increased in the last decades (Bohman 1997, 1998, 2000).

As Cass Sunstein (2002, 2005) puts it, deliberative democracy should not be judged as superior to other forms of government such as autocracy or monarchy. The fact is that before the empire of risk, deliberative democracy allows consented answers which unite the citizenship. However Sunstein agrees that democracies run serious risk when populism emerges. At a closer look, under an authoritarian government most probably disasters might suddenly efface the community because information is strictly monopolized by the State. In the democratic nations, information flourishes everywhere and is daily exchanged by the citizens. This means that democracies are best prepared against disasters but the State should not cede to the populist claims. According to Sunstein, populism centers on "the risk neglect", which means the emotional distortions and prejudices that normally condition the decision-making process. Officials should listen to the experts who are trained to deal objectively with risk. The popular opinion, in the democratic context, succumbs to fear and panic, demanding more than the State can meet. As a result of this, the lay-citizens echo risks which have minor impacts on society while other more dangerous risks are glossed over. Sunstein's preliminary remarks deal with the question of why people are frightened, or, as an alternative, why people feel safe when they should feel fear. Sunstein first examines the role played by rationality in the process of dread and its consequent relationship with democracy. One of the aspects that distinguishes deliberative democracy from other forms of government is the presence of debate, common debate, involving a majority of different classes and citizens. To put this in other terms, deliberative democracy disposes society to be prepared before disasters, or better equipped than demagogues or authoritarian governments. In deliberative democracies, external risks or situations may be commonly discussed with cooperation to reach new alternatives or instruments. The theorists of deliberative democracy suggest that – unlike totalitarian regimes – the sense of freedom which facilitates the debate allows citizens to be actively engaged in evaluating and correcting the next step to mitigate the negative effects of potential disasters.

Undoubtedly, spectators over-stimulated by news of terrorism, anxiety and fear would be more sensitive to accept torture as a valid alternative to strengthen security. In the days of terror, as Altheide amply validated, people are receptive to embrace totalitarian discourses which jettison the obtained individual rights of some minorities or at best adopt radicalized behavior (Altheide 1997, 2002, 2003, 2006). As with the previous argument, in the book *Television and Terror* A. Hoskins & B. O'Loughlin (2009) question the existent political allegories replicated by the media (CNN to be exact) regarding the war on terror and the ever-increasing sentiment of insecurity felt by people in urban areas. They coin the term "democratic imperialism" as the widely disseminated political discourse aimed at creating a democratic alliance (against terrorism) in order for homeland security to be tightened. This gradually ushers nations into an "assertive multilateralism" which seeks to strengthen the actions (as Habermas wanted) of international institutions such as the UN or NATO. Departing from the belief that this architecture can be orchestrated through a "modulated fear", promoted and amplified by the media, the (liberal) state reserves the use of more repressive forms of coaction at its discretion. Though entertained by the show of terrorism, as Hoskins & O'Loughlin claim, the audience is carefully surveilled and spied on by its government. The "rhetoric of fear" connects directly with many different events, recreating a climate of anxiety which domesticates the dissension. Ultimately, Douglas Kellner (2015) examines the impacts of media culture in the struggle Western democracies pursue against terrorism. We live in the days of a "techno-capitalism", which moves us toward the society of the spectacle. In perspective, the goals of the media are not strictly associated with the truth, but to forge a dependent audience that can be bombarded by fake news. For Kellner, like many other academicians we have already discussed in earlier chapters, 9/11 inaugurated a new epoch where the disaster is spectacularized, commoditized to be exchanged to a news-thirsty global public. While the war on terror increases the sentiment of vulnerability of lay-citizens, the media maximizes its profits. Nonetheless, the performance of media or celebrities in these much deeper issues does not resolve why democratic governments appeal to torture to gather vital information in their crusade against the Islamic State.

Is torture enough?

As Fritz Allhoff (2003) mentioned, the tragic events of the World Trade Center animated the discussion revolving around torture in the US. Some national polls say that some Americans ethically endorse torture as a valid mechanism of protection whereas the majority rejects it. Allhoff is heavily convinced that torture works in some cases and conditions. According to his stance, though the US is a consolidated democracy, it is naïve to think that vital information coming from torture or clandestine detention will not be accepted by the authorities. Allhoff declares himself more interested in discussing the morality of torture instead of its legality. While the former is subject to universal values which define us as humans, the latter can be politically manipulated, reaching a temporal existence. The discrepancies between the deontological and the utilitarian approach relate to the position each one adopts to define torture. The utilitarian perspective is based on "the ticking time-bomb" paradigm which sustains that torture may be positively employed to save life before an imminent attack. One of the limitations of this posture is that the concept of immediacy seems to be very hard to define precisely. Allhoff recognizes there are many problems around "utilitarianism" that are duly documented and espoused to public opinion over recent years. Rather, the deontological position signals the prohibition of torture in almost all cases and circumstances. Taking their cues from Kant, the supporters of this theory emphasize that torture should be viewed as a failure to respect the integrity of the "Other" and their dignity. Allhoff believes that torture does not violate any right when the intention of criminals or terrorists is to kill others. Even torture is ethically permissible when important information can be retrieved and this helps to save lives, or prevents a major threat to society. Stritzke et al. (2005) call to attention that torture is widely denounced by activists, NGOs and underdeveloped governments while the central powers defend it as part of a valid toolkit to tighten homeland security. What we know is that torture and terrorism seem to be two evils, which work together. As I noted in earlier studies, terrorism evinces a dialectics of hate, which combines a torture-sponsored state with the fight of insurgents who instrumentalize innocent others to achieve their own goals. No matter the side or the band, both sides share some commonalities such as "the indifference for the other's suffering", the need to expand instrumentalization, which means using others to fulfill my own goals, and the sudden blow as a form of creating political instability (Korstanje 2017). It is not surprising, if the reader follows this argument, that torture would persist over the years in the hub of operation of military forces, regardless of the political organization of the nation. Of course, democratic nations may develop some resources and background to deal temporarily with violence while authoritarian governments (like the crimes committed by Juntas in Latin America) are prone to extend violence unleashed indefinitely (Korstanje 2017). In consonance, Alex Bellamy (2009) doubles the bet arguing that liberal states play an ambiguous role because they foster torture for some specific cases as terrorism, while other countries are strictly sanctioned or discouraged to use it. As he brilliantly comments, history teaches us that torture generates counterproductive effects, often inversely proportional to the original ends. The IRA tripled its attacks against Britons after members of the IRA were tortured in the same way that Al-Qaeda never stopped the violence despite the existence of Abu Ghraib (Iraq). Though linked, experts should discern the use of torture not by the reasons behind it but by the potential effects.

In an innovative essay-review, Matthew Hannah (2006) draws the geopolitical contours of terrorism and torture, placing torture as a biopolitical technique that compensates the idea of exemption initially formulated by Agamben. Terrorism is not a danger for peoples, but for the topological presuppositions encapsulated in the discourses of power-knowledge that territorialize the adscriptions of citizens to the soil. Claudia Card (2010) interrogates the aftermaths of 9/11 through the lens of "atrocity theory". Like Ignatieff, she accepts the doctrine of "lesser evil", juxtaposing the evils from justified violent crimes. The atrocity paradigm refers to a recently formulated theory intended to outline that evils are naturally unbearable harms perpetrated by culpable evildoers. The proponents of the atrocity paradigm say that evildoers necessitate being wicked persons and there are no possibilities to reason or dialogue with them. In the same line as Zimbardo, she holds the thesis that evildoers should not be mythologized or demonized; they are humans like us. The individual dispositions are subject to the cultural background in which case we must divide evils from "lesser wrongs".

In the midst of a misleading debate, one might question to what extent the sense of terror paralyzes the ethical conception of "the being". This was the main theme of Richard Bernstein (2002) when this tragic event surprised him while typesetting the end of *Radical Evil: A Philosophical Interrogation*. In this text, Bernstein digested the nature of Auschwitz and the crimes on innocent civilians perpetrated by Nazi Germany. The rise of terrorism prompted Bernstein to start a new project he named *The Abuse of Evil: The Corruption of Politics and Religion since 9/11*. For readers and students to understand the development of Bernstein, I must add, it is important not to lose sight of the performance of religion and nationalism in the configuration of extreme (radicalized) narratives aimed at dehumanizing the alterity. The figure of suffering has a biblical connotation. To set an example, Lucifer commanded a rebellion against God and was castigated accordingly. From that moment onwards, he emulated the archetype of evilness. For these founding myths, the omnipotence of God is pitted against the greed of this fallen Angel. Philosophers and theologians devoted considerable efforts in elucidating the nature of evil. But for humans, things are completely different. Evil should be understood as a social construct, based on the triviality of ethics, or in terms of Arendt as the banality of critical thinking. Unless otherwise resolved, the evil aims to trivialize the essence of humanity in so far as this is situated as the main chosen tactic of totalitarian governments. Although the history of the twentieth century is fraught with slaughters and genocides, the 9/11 attacks and the attention of media corporations emptied the symbolic core of evilness, posing Osama Bin Laden and Saddam Hussein as the villains of Western civilization. The discourse clearly highlighted the lack of tolerance of religion and particularly the Muslim faith in respecting democracy and the Western lifestyle. From their inception, religion as well as politics were framed to forge a collective well-being. The success of terrorism hinges on its ability to tergiversate the nature of religion and politics, paving the way for the emergence of a mythical conflagration between good and evil. In this mayhem, philosophy should discuss critically the need for people to adopt universal absolutes as shelters, which exhibits fertile ground for terrorism. In times of anxiety, fear and uncertainty, Bernstein writes, philosophy acts as a critical conduit preserving democracy from totalitarian claws. Equally important, pragmatism is pivotal in dissociating practice from belief to avoid radical reactions that may place democracy in jeopardy. The recent conflicts in the Middle East, adjoined to 9/11, lead people to reject their flexibility respecting the non-Western Other. One of the paradoxes of terrorism indicates that both parts feel God is supporting their causes, interpreting the doctrinal texts according to their convictions, interests or prejudices. If God is on my side, torture seems to be a lesser evil I am divinely allowed to use when the opportunity arises. The process of secularization facilitated the creation of multiple realities that should be traced back to the roots of political conflict. Evil not only takes different shapes but also legitimate personal interests.

> The expression relating to war against terror is pretty deceitful. Terror appears not to be an enemy system of tactics and strategies aimed by many collectives with diverse ends on mind. However, all those who are eager to label enemies as wicked or part of the axis of evil are the reluctance to know why the rest of World get on well with terrorism...
>
> (Bernstein 2005: 102–103)

To cut the story shorter, torture is not an effect or the immediate result of terrorism, but the evident sign of the decomposition of politics in our postmodern times. Here Bernstein exerts the most caustic critique on contemporary democracy since it is a ritual performance that takes place every four years. However, we must understand democracy as something other than a ritual but an egalitarian lifestyle. The politics are being corrupted through the introduction of fear as a paralyzing narrative that enhances the workforce's loyalties to its leaders (Bernstein 2005).

One of the most interesting argumentations against torture we have read should be traced to the book edited by Sharon M. Kaye (2007) *Lost and Philosophy:*

The Island Has its Reasons. One of the compiled chapters authored by the young researcher Scott Parker (2007) brings reflection on the inefficiency of torture to reveal key information in view of the immediacy of the next terrorist attack. Torture simply does not suffice in the fight against radicalized cells because innocent people – before the extreme tribulations and suffering – will say anything for the pain to stop, and guilty persons stoically resist the torture or blame others. Torture, without any doubt, creates a spiral of terror and violence which is very hard to halt in a no-so-distant future. Through torture there is an escalation of violence which, legally regulated or not, does not resolve the problem (Parker 2007). This reveals brilliantly what are the most common paradoxes of violence systematically applied to terrorist suspects.

Conclusion

To conclude, though this theme should be continued in a future approach, liberal democracies operate with double standards. On one hand, they prohibit the use of violence and torture to vulnerate civilians. This is the case of Jimmy Carter's denunciation of human rights violations in Chile and Argentina. The Juntas, frightened by the arrival of communism, started a death-machine oriented to exert a radical repressive reaction against subversion. Immediately, international public opinion energetically condemned these Latin American governments with the focus on the democratic ideals of the republic. The analysts and specialists who studied issues of this caliber systematically emphasized the undemocratic nature of the Juntas. Argentina, Chile and Uruguay, by their diagnoses, were historically subject to intermittent democracies alternating coups with weak civic governments. No less true was that military forces took power, adducing the needs for liberal order and stability which were jeopardized by the advance of communist parties and cells (Dinges 2005). In this perspective, the acts of torturing suspected prisoners, once the Juntas set the pace for Alfonsin's presidency, were cataloged as crimes against humanity. Of course, the vulnerability of the victims and the advantages of the military that monopolized the material and financial resources of the State awakened condemnation by the most representative democracies in the world.

Nonetheless, if 9/11 proved something it was that even consolidated democracies may fall into abuses of all types and even torture. The discussion in the US did not focus on these grim, undemocratic cultures geographically located in Latin America; the monster now was knocking at the doors of Paradise. The US failed to give a credible answer to the allegations of torture in Supermax prisons at the same time as parliament ratified these spaces of detention as secure and efficient in the war on terror. This chapter has brought the problem of torture in the most mature democracies into the foreground. Through the reading and discussion of this chapter, we found that torture results from the instrumental nature of the liberal nation-state. The discourse of instrumental reason, which is enrooted in a modern rationale, rests on the logic of means and ends. The main ideological core of modern instrumentality assumes that the body should be saved when the part can be suppressed. This is exactly the medical metaphor. When doctors judge that an organ of the body should be extracted to save the patient's life, they do not hesitate to do it. The same applies to terrorism, and ethnic minorities are often targeted as potential enemies of democracy, ranging from Muslim believers, many of them naturalized or second-generation native citizens, to asylum seekers, refugees and migrants who experience the dark side of American hospitality in Trump's presidency. Whatever the answer may be, instrumentality seems to be the key indicator of modern democracies and terrorism, a point which deserves to be discussed in the next chapters. This raises some concerns, such as: is technology part of the problem or the solution? Are Americans sensitive to end-of-days theories? What are the connections between technology and the end of the world?

References

Albrecht-Meylahn, J. (2013). Divine violence as auto-deconstruction: The Christ-event as an act of traversing the neo-liberal fantasy. *International Journal of Zizek Studies*, *7*(2), 1–19.

Alexander, Y. (Ed.). (1976). *International Terrorism: National, Regional, and Global Perspectives*. New York, Praeger.

Allhoff, F. (2003). Terrorism and torture. *International Journal of Applied Philosophy*, *17*(1), 121–134.

Altheide, D. L. (1997). The news media, the problem frame, and the production of fear. *The Sociological Quarterly*, *38*(4), 647–668.

Altheide, D. (2003). Notes towards a politics of fear. *Journal for Crime, Conflict and the Media*, *1*(1), 37–54.

Altheide, D. L. (2002). *Creating Fear: News and the Construction of Crisis*. Transaction Publishers.

Altheide, D. L. (2006). Terrorism and the politics of fear. *Cultural Studies ↔ Critical Methodologies*, *6*(4), 415–439.

Appiah, A. (2001). "Grounding Human Rights". In *Human Rights as Politics and Idolatry*, M. Ignatieff (ed.). New Jersey, Princeton University Press, 101–116.

Bellamy, A. J. (2009). "Torture, Terrorism, and the Moral Prohibition on Killing Non-combatants". In *Terrorism and Torture: An Interdisciplinary Perspective*, W. G. K. Stritzke, S. Lewandowsky, D. Denemark, J. Clare, & F. Morgan (eds). New York, Cambridge University Press, 18–43.

Bernstein, R. J. (2002). *Radical Evil: A Philosophical Interrogation*. Cambridge, Polity Press.

Bernstein, R. J. (2005). *The Abuse of Evil: The Corruption of Politics and Religion since 9/11* (Vol. 19). Cambridge, Polity Press.

Bohman, J. (Ed.). (1997). *Deliberative Democracy: Essays on Reason and Politics*. Cambridge, MIT Press.

Bohman, J. (1998). Survey article: The coming of age of deliberative democracy. *Journal of Political Philosophy*, *6*(4), 400–425.

Bohman, J. (2000). *Public Deliberation: Pluralism, Complexity, and Democracy*. Cambridge, MIT Press.

Borradori, G. (2013). *Philosophy in a Time of Terror: Dialogues with Jurgen Habermas and Jacques Derrida*. Chicago, University of Chicago Press.

Brown, W. (2008). "Power after Foucault". In *The Oxford Handbook of Political Theory*, J. Dryzek, B. Honig, & A. Phillips (eds). Oxford, Oxford University Press, 65–84.

Butler, R. E. (1976). Terrorism in Latin America. *International Terrorism*. New York, Praeger.

Calveiro, P. (2012). *Violencias de Estado: la guerra antiterrorista y la guerra contra el crimen como medios de control global.* (*Violences of States: The War on Terror and the Fight against Local Crime as Disciplinary Means of Global Control*). Buenos Aires, Siglo XXI.

Card, C. (2010). *Confronting Evils: Terrorism, Torture, Genocide.* Cambridge, Cambridge University Press.

Castel, R. (1997). *La Metamorfosis de la Cuestión social. Una Crónica del salariado.* (*The Metamorphosis of Social Question*). Buenos Aires, Paidos.

Castoriadis, C. (2006). *Lo Que Hace a Grecia. De Homero a Heráclito.* (*What Makes Greece*). Buenos Aires, Fondo de Cultura Económica.

Chappell, Z. (2012). *Deliberative Democracy: A Critical Introduction.* New York, Palgrave Macmillan.

Chomsky, N. (2011). *9–11: Was There an Alternative?* New York, Seven Stories Press.

Chomsky, N. (2015). *Culture of Terrorism.* New York, Haymarket Books.

Cistelecan, A. (2012). Which critique of human rights? Evaluating the postcolonial and the post-Althusserian alternatives. *Wronging Rights? Philosophical Challenges for Human Rights, 1*, 1.

Cronin, A. K. (2011). *How Terrorism Ends: Understanding the Decline of Terrorist Campaigns.* Princeton, Princeton University Press.

Dinges, J. (2005). *The Condor Years: How Pinochet and His Allies Brought Terrorism to Three Continents.* New York, The New Press.

Dryzek, J. S. (1999). Transnational democracy. *Journal of Political Philosophy, 7*(1), 30–51.

Eid, M. (2014). "Terroredia: Exchanging Terrorism Oxygen". In *Exchanging Terrorism Oxygen for Media Airwaves: The Age of Terroredia.* Hershey, IGI Global.

Foucault, M. (1982). The subject and power. *Critical Inquiry, 8*(4), 777–795.

Foucault, M. (2003). *"Society Must Be Defended": Lectures at the Collège de France, 1975–1976* (Vol. 1). London, Macmillan.

Foucault, M. (2007). *"Security, Territory, Population": Lectures at the Collège de France, 1977–78.* New York, Springer.

Foucault, M. (2012). *The Archaeology of Knowledge.* Vintage.

Freeman, M. (2011). *Human Rights. An Interdisciplinary Approach.* Cambridge, Polity Press.

Gutmann, A. (2001). "Introduction". In *Human Rights as Politics and Idolatry*, M. Ignatieff (ed.). New Jersey, Princeton University Press, vii–xxviii.

Habermas, J. (2015). Democracy in Europe: Why the development of the EU into a transnational democracy is necessary and how it is possible. *European Law Journal, 21*(4), 546–557.

Hannah, M. (2006). Torture and the ticking bomb: The war on terrorism as a geographical imagination of power/knowledge. *Annals of the Association of American Geographers, 96*(3), 622–640.

Hardt, M., & Negri, A. (2001). *Empire.* Cambridge, Harvard University Press.

Hobbes, T. (2006). *Leviathan.* London, A&C Black.

Hoskins, A., & O'Loughlin, B. (2009). *Television and Terror: Conflicting Times and the Crisis of New Discourse.* New York, Palgrave Macmillan.

Howie, L. (2012). *Witnesses to Terror: Understanding the Meanings and Consequences of Terrorism*. New York, Springer.

Ignatieff, M. (2001). *Human Rights as Politics and Idolatry*. Princeton, Princeton University Press.

Ignatieff, M. (2005). "Introduction: American Exceptionalism and Human Rights". In *American Exceptionalism and Human Rights,* M. Ignatieff (ed.). Princeton, Princeton University Press.

Ignatieff, M. (2013). *The Lesser Evil: Political Ethics in an Age of Terror*. Princeton, Princeton University Press.

Jeffreys, D. (2013). *Spirituality in Dark Places: The Ethics of Solitary Confinement*. New York, Palgrave Macmillan.

Kaye, S. (2007). *Lost and Philosophy: The Island Has Its Reasons*. New York, Wiley-Blackwell.

Kellner, D. (2015). *Media Spectacle and the Crisis of Democracy: Terrorism, War, and Election Battles*. New York, Routledge.

Korstanje, M. E. (2013). Empire and democracy: A critical reading of Michael Ignatieff. *Nómadas, 38*(1), 1–19.

Korstanje, M. E. (2017). *Terrorism, Tourism and the End of Hospitality in the West*. New York, Springer Nature.

Korstanje, M. (2018). *The Mobilities Paradox: A Critical Analysis*. Cheltenham, Edward Elgar.

Lyon, D., & Bauman, Z. (2013). *Liquid Surveillance: A Conversation*. John Wiley & Sons.

Marx, K. (1983). *The Portable Karl Marx*. New York, Penguin Group USA.

Morefield, J. (2008). Empire, tragedy, and the liberal state in the writings of Niall Ferguson and Michael Ignatieff. *Theory & Event, 11*(3), 1–20.

Parker, S. (2007). "Torture Souls". In *Lost and Philosophy: The Island Has Its Reasons*, S. Kaye (ed.). New York, Wiley-Blackwell, 148–158.

Robin, C. (2004). *Fear: The History of a Political Idea*. Oxford, Oxford University Press.

Simon, J. (2007). *Governing through Crime: How the War on Crime Transformed American Democracy and Created a Culture of Fear*. Oxford, Oxford University Press.

Skoll, G. R. (2016). *Globalization of American Fear Culture: The Empire in the Twenty-First Century*. Springer.

Souter, J. (2009). Humanity, suffering and victimhood: A defence of human rights pragmatism. *Politics, 29*(1), 45–52.

Soyinka, W. (2005). *The Climate of Fear: The Quest for Dignity in a Dehumanized World*. New York, Random House.

Stampnitzky, L. (2013). *Disciplining Terror: How Experts Invented "Terrorism"*. Cambridge, Cambridge University Press.

Stritzke, W. G., Lewandowsky, S., Denmark, D., Clare, J., & Morgan, F. (2005). *Terrorism and Torture*. Cambridge, Cambridge University Press.

Sunstein, C. R. (2002). *Risk and Reason: Safety, Law, and the Environment*. Cambridge University Press.

Sunstein, C. R. (2005). *Laws of Fear: Beyond the Precautionary Principle* (Vol. 6). Cambridge, Cambridge University Press.

Zimbardo, P. (2007). *The Lucifer Effects: How Good People Turn Evil*. New York, Random House.

5 The dark side of technologies

The industry of fear and the apocalypse

Introduction

The zombie virus has finally gone viral. The cable television series *The Walking Dead* had seven million "likes" on Facebook and 300,000 Twitter followers as of March 2012 (Lazar 2012). Vampires, if anything, show even larger numbers with an estimated 32 million Facebook likes (Graphs.net 2013). Other undead populate the web and popular culture. The undead in various forms may not inhabit the earth, but they proliferate and reproduce in electronic form along with print media. We suggest that this phenomenon, the popularity of undead motifs, does not arise from especially clever marketing strategies, although they play a role, they would find less success if it did not resonate with a form of public consciousness, or more accurately, unconsciousness. The undead represents a postmodern sensibility. This sensibility reeks of decay.

> "[It is a] 'degraded' landscape of schlock and kitsch, of TV series and Readers' Digest culture, of advertising and motels, of the late show and the grade-B Hollywood film, of so-called para literature with its airport paperback categories of the gothic and the romance, the popular biography, the murder mystery and science-fiction or fantasy novel: materials they no longer simply 'quote', as a Joyce or a Mahler might have done, but incorporate into their very substance" (Jameson 1991:55).

Jameson goes on to assert that postmodern culture represents a political unconscious. That is, people experience a postmodern political economy, but they lack the wherewithal to express it in articulate discourse, in political terms. In a similar vein, the arts have represented, in what might be considered prescient ways, what has already begun in the economic, political, and social structures, but has not yet appeared in explicit terms. Somewhat arbitrarily stated, modernism began in the mid-nineteenth century. Charles Baudelaire called it modernité. Accordingly, modernism depicted the ephemeral, fleeting, ever-changing nature of industrialized urbanism. Even then Baudelaire's fascination with the macabre and his admiration for the work of Edgar Allan Poe portended a certain connection between death, decay, and the advent of modern culture.

(Skoll & Korstanje 2014: 11)

Doubtless the obsession of modern man to control nuclear energy leads towards unseen risks such as accidents or potential nuclear war. As U. Beck (1992) puts it, one of the paradoxes of modernity started with the Chernobyl accident, where the same technology disposed to make this world a safer place, may impact negatively on the environment. Beck acknowledges that while some risks are successfully mitigated by the action of technology and rational technique, new ones emerge. What is more important, the state of disaster results from the excess of modernity (Beck 1992). Jacques Ellul interrogated philosophically the negative effects of technology in modern society as not only undermining critical thinking but also controlling – if not commoditizing – other lay-citizens (Ellul 1962, 1992). Andrew Feenberg calls attention to technology, à la Marcuse, as a form of new rationalization, which means control over the rank-and-file worker. Technology controls human bodies to enhance the means of production but tries to domesticate nature through the capitalist wage system. While machines operate to make our lives safer, Western rationality imposes over other voices and cosmologies (Feenberg, 1995).

Echoing these above-mentioned concerns, Zygmunt Bauman (2013) elicits an interesting critique on the ideological hegemony of machine in the liquid society. From his viewpoint, society cultivates a climate of emerging fears, which mediate between peoples and their institutions, and of course it runs the risk that real disasters cannot be prevented. The natural barriers that often alert us to the looming risks are undermined in view of a spectacle, oriented to entertain the masses. In consequence, when the real disasters take place society has lost its immunity. Jointly with David Lyon, Bauman, in the book *Liquid Surveillance* (2013), calls attention to the fact that the growing process of securitization corresponds to a need for status and social distinction. 9/11 did not create but accelerated the conditions of exploitation of capitalism over the workforce. The digital cameras and surveillance technology were disposed not only to protect the interests of the ruling elite, but also to show others they are, after all, outstanding, special and have the necessary resources to remain isolated from the rest of society.

Bauman & Lyon echo Arendt's contributions, above all her approach on the banality of evil, which means the triumph of bureaucratic reason. The authors continue the Foucauldian insights on the panoptic. However, as they put it, this concept of the panoptic was notably changed. Now, a large, global audience watches a few people in a reality show. Another interesting point in this discussion, posited by Bauman and Lyon, consists in the dichotomies or the contradiction of mobilities. While first-world tourists are legally authorized to travel worldwide, states scrutinize and tighten their borderlands to thousands of asylum seekers, forced migrants and refugees. This humanitarian crisis not only is inevitable, but also seems to be the result of the sentiment of indifference Western governments have. Whether over centuries the alien was frightening, globalization made us fear our neighbors. In view of this, the use of surveillance technologies has two important functions. On one hand, citizens need to feel secure in an ever-changing world that places their ontological security in jeopardy. On the other hand, by buying this technology they join the "chosen people", exactly those who have the purchasing power to contract private security. Whatever the point may be, the process of securitization reveals a philosophical dilemma: the more technology is used to make the homeland a secure place, the more insecurity citizens really experience. To cut a long story short, technology – as discussed – serves as a sign of status that separates classes and citizens, reinforcing the disparities and inequalities of capitalism.

The goals of this chapter are manifold but they can be synthesized as follows:

- Learning more about the connection between technology and civilization.
- Discussing critically the point of convergence of apocalypse days and the theory of evolution.
- Laying the foundations for a new debate regarding the role of technology in the apocalypse-related narratives.

In the next section we shall place the concept of law and order under a critical lens, alternating different views and theories to try to construct a bridge between Liberalism and Marxism.

Law and order

One of the questions that concerned philosophers and social scientists was how society is united. Thomas Hobbes (2006) envisaged the social bond as based on two powerful but contrasting tendencies: the appetite for the property of others, and the needs of personal safety. Hobbes goes on to say that one might fight to gain further wealth competing with others, but sooner or later, the concept of protection prevails. To avoid the war of all against all, people do confer to a third party, Leviathan (State), the monopoly of force. Therefore, in Hobbesian theory, human beings are prone to develop a lasting peace. In contrast, Jean Jacques Rousseau (1997) argued that human beings corrupt themselves when they abandon the state of nature as given by God. The societal order gives to the person an illusory view of reality. The division of labor, accompanied by all ideological mechanisms of indoctrination, leads the individual savage mind to the covenant. Certainly, the conflict we in our societies observe today results from this frustration and not vice-versa. The Rousseau legacy gave an all-encompassing idea of how a group is formed, and it paves the way for the contributions of another French scholar, Emile Durkheim.

Durkheim (1893, 1895, 1912) turned to Rousseau's concept of the volonté générale for a conception of social solidarity that did not depend on the atomistic individualism of liberal economics, today neoliberal economies, nor on the Hobbesian Leviathan. Durkheim recognized that contractual relations among people, even those that ostensibly founded the State, depended on pre-existing norms and shared values. Here, Durkheim inserted the conscience collective as a way to talk about how humans carry on their intercourse on a quotidian basis – that is neither through formal contract nor coercion. People collectively create their normative worlds, which they then project onto various organizational and symbolic entities, hence arise churches, states, and religions.

(Skoll & Korstanje 2014: 12)

Sandra Gasparini (2015) evokes the figure of the specter and the zombies as two forms of control which are often associated with post-apocalyptic landscapes. Basically, as she notes, the ghost interrogates the subjectivity of the human soul while the zombie appeals to social order, which in a Hobbesian term leads towards a collective fear.

The liberal alternative

While Durkheim sought an explanation for how a society is possible in collective explanations, the liberal alternative stays close to individualism. Instead of a coercive leviathan as in Hobbes, an apparently softer version was articulated by John Locke (1689). In the Lockean version of the social contract, the State acts as an arbiter, whose main role is to settle disputes – a sort of general arbitration panel. John Rawls was one of the main exponents of this liberal tradition.

Based on his theory of justice, John Rawls enumerates five forms of political organizations. In his 1999 The Law of Peoples he focuses on some worries relating to a utopia within the framework of a democratic constitutional society. Starting from this premise, Rawls explains not only why some nations fail in this process but also the idea that motivates the law of peoples in modern democracies. The first, he adds, is one of the evils of human history, unjust war and political submission. The second signals to social policies to efface injustices. Rawls' five subtypes of political organizations are: a) reasonable law, b) decent people, c) outlaw states, d) societies burdened by unfavorable conditions and e) benevolent absolutism …

… What is important to discuss in the Rawls's theory is to what an extent the liberality he proposes creates asymmetries or inequalities. He proposes the veil of ignorance in establishing a constitution in which the founders of a polity will not know where they will stand in the social and political-economic hierarchy. By this means, he argues, the constitution will ensure a measure of equality and equity, because no one wants to be at the bottom and if they are, wants the ability to rise.

(Skoll & Korstanje 2014: 12–13)

As Wolfe (1983) brilliantly observed, the idea of the end of the world suggests three significant axioms which are related not only to the power of technology but the emergence of chaos into the subjective world, as well as a mythical battle between evil and good, where the human character prevails over destiny. In fact, "the assurance of racial survival despite the most overwhelming odds – a kind of denial of death" (Wolfe 1983: 6) is determined by the heroic behavior in an individual dimension. This means the man, not Gods, seems to be the architect of his own destiny.

The end of the world

This section is an extract from Skoll & Korstanje (2014: 14–17).

Damian Thompson (1998) says that millenarianism refers to the belief that civilization is facing the end of the world as a result of sins or other moral failings. This situation often is depicted as the encounter of good versus evil. After the latter is defeated, humankind is favored by a thousand years of prosperity and happiness. Thompson reminds us millenarianism is not a prerequisite of religious life. Moreover secular organizations in political life appeal to millenarianism in particular contexts. If the world advances inexorably to its own self-destruction, the New Testament reveals that a select group of persons will be saved. This, Thompson adds, leads to narcissism and maniacal grandiosity. It is not surprising this type of attitude favors mass suicide and similar pathologies; Jonestown comes to mind. Pertinent here is the connection of millenarianism with the economy. From Mircea Eliade on, anthropologists turned their attention to apocalypses as a projection of the economic cycle of the soil. Like a calendar that leaves some days for destruction, apocalypses appeal to a much broader discourse of disasters. Signs of terror accompany millenarianism and point to the end as a product of self-corruption. The root of bottom days is the notion that sin must be expiated. Disasters may be equated to having a bad harvest. In many cultures, this is commemorated by a New Year. The bottom-days theorist realized that after the economy encountered problems, millenarianism surfaced as an expiatory ritual of restitution (Thompson 1998).

Almost all cultures have tales about an exemplary center, an immemorial life where human beings and God coexisted in harmony. This paradise, a term coined by Eastern religions, has prevented access by humans, the founding parents having committed a crime, sin or other norm violation. From that day on, civilizations try to regain the primitive state of nature where suffering did not take place.

Thompson argues that if humankind lives in accordance to God's precepts, the cycle of decadence (downfall) may be deterred. For some reasons, the bottom-days appears whenever the economic order is in danger, or in the process of basic change, like Zoroaster or the Book of Daniel that was written in times of uncertainty, war, and chaos. Apocalyptic literature seems to be associated with a rupture between two orders or cosmologies. Accompanied by an awful and shocking vision, prophets serve as mediators between the present and the shadows of future. The state of emergency that apocalyptic related literature generates is, for some analysts, a fertile source for messianic politicians. Thompson emphasizes the process of isolation that millenarian groups suffer. Since these attitudes break contact with outsiders, and reduce the number of social bonds, rendering members more vulnerable to psychological manipulation. The characteristics of apocalypses and bottom-days prophecies are detailed below:

- The message was given by God(s) or Angel(s) and defies the prevailing political authority.
- The destruction of oppressors not only is imminent but an irreversible decision of gods.

- It appeals to a much broader sentiment of exception that sometimes leads to narcissism.
- The vision is revealed to a community, whose morality is set above that of other groups.
- The end of the world starts with a fight between two opposing forces – good and evil.
- There is a tendency to combine hope, linked to faith, with goodness and fear to evilness.
- The proximity of the great disaster triggers a mass migration, abandoning the old home.
- The calendar takes a month (28 days), to be divided by the divined number, 4. This gives 7, the days of the week. Thompson agrees though this has been imported by Judaism (known as the theory of great Week), it has been practiced in Sumer and other neighboring communities. Following this argument, bottom days would be a question of working time.
- The narrative is accompanied by disasters, which are caused by the moral decline of humankind.

The allegories used through these tales are emulated and projected to daily life. For example, John from Patmos had the vision of a great beast, a term which later came to be associated with Stalin, Hitler, Kissinger and even the Pope. The malleable nature of apocalypse makes the discourse perdurable in all epochs. In his book, *The End of Time*, Thompson demonstrates that the cycles of time are determined by the economy, but when some problem presages an upcoming crisis, the idea of the end of time serves to recuperate the status quo by renovating the pride and trust for the community all – a reaffirmation of its collective conscience. In view of this, politicians whose administrations are not successful are prone to manipulate apocalyptic literature to cause two effects: fear and hope. Based on a state of rivalry between what is considered the evil and good, or the essence of corruption, the political order tends to be perpetuated. Following this explanation, it is not surprising that the Mayan prophesies of the bottom days in 2012 would be a result of the financial crisis that whipped the world in 2007. Although technically this event has a clear diagnosis, terrorism, natural disasters, global warming, and other traumatic events were interpreted as a sign of the end. Of course, a discourse of this caliber means that no real solutions for real problems are needed. Not coincidentally, the citizenry is domesticated by the introduction of daunting news. People feel as if nothing can be done simply because the end is imminent. This discourse gives elites more legitimacy and undermines the critique inside the society.

Every end of millennium represents for human beings a new structuration of their beliefs, their production, forms of consumption, and even their hierarchical lines of authority. Millenarianism can be defined as a moral movement whose aims are intended to eradicate the sin of the human heart; at the same time, it offers a new pattern of behavior rooted in brotherhood, love, and cooperation. As privileged witnesses to the beginning of a new millennium, current scholars play a pivotal role in the understanding of these types of radical changes for humanity, although this topic has not gained considerable attention from social sciences for now. Some members of societies invent their own the eschatology or the bottom days. Sometimes they are supported by previous polarized beliefs that emphasize the in-group as a sign of good while the others are depicted as a representation of evil. For that reason, the present review is twofold. It refers to the different anthropological waves that focused on millenarianism, and it reads a social discourse which can be examined to reconstruct the life of writers who prophesized the end of the world.

Norman Cohn (1996) surmises that Zoroaster (c. 1000 and 1500 BCE) was the first prophet who addressed the apocalypses which symbolize the final fight between Asha's forces (order) and Druj (chaos). Evidently, Zoroaster's life was marked by a time of turbulence and conflicts. Zoroaster was experiencing the invasion of new Indo-Aryan neighbors. Most likely, an Iranian invasion pushed him and their relatives to escape to other remote zones. Under the dichotomy of domination or liberation, Zoroaster projects in his texts the ambivalence involved his deeper emotions. After all, the theory of apocalypses is not too strange. The message of apocalypses is aimed at articulating a devastation resulting from human corruption. Suffering, redemption, and the rediscovery of tragedy are parts of Zoroastrianism interconnected in Christianity. One of the relevant aspects predominate at the end of the world seems to be the injustice and despair.

Bernard McGinn (1996) examined the consequences of the fall of the Roman Empire. According to McGinn, Medieval Christianity dissociated the conceptualization of civilization from Christianization. These societies were in part fragmented and redeemed their disputes in civil wars on behalf of religion. The belief in the last judgment, encouraged by theologians, led to the need to forecast when and how the world would end. The theory of apocalypses and end of the world converge on hopes for a new life and a thousand years of peace within the context of terror in a realm ripe for total destruction. McGinn addresses the legends giving origin to the theses of Beda, Gregorio Magno, and Saint Columbanus (three scholars who sustained the idea that a Sacred Roman Empire should be erected following Catholic principles). The figure of Christ as a king of kings was, of course, functional to the contemporary political power of a nobility. The historical frame ranges from 400 to 1000 CE, which corresponded to a new sense of Christianity based on the meaning of Christ as a king.

Krishan Kumar (1996) emphasizes the apocalypse as a figure of utopia. In this valuable project, Kumar catches a glimpse of the different sentiments predominating in European societies during the tenth century. He compares the fear and expectation of the Middle Ages with our postmodern gaze, arguing that we are witness to a decline of hope. The secularization of disaster has been created by humans so as to experience the terror of an imminent end. Unlike the Middle Ages, the lack of hope puts humanity in a difficult position. The tenth century was not the time of unmitigated terror, even though there were some natural tensions. Saint Augustine and the Venerable Bede had hopefully anticipated the end of the world in 999. Generally, the Roman Catholic Church considered prophecies a form of heresy, but with certain toleration. The Protestant Reformation would reinvent the apocalyptic traditions continuing with the exegetic and historical revisions linking directly the figure of the Antichrist with the Pope. In this vein, professor Kumar accepts that the mayhem brought by the advent of the tenth century was accompanied by a renaissance of culture and economic growth. The bottom-day destruction by imposition of blood, pestilence, and violence was certainly associated with a moral need for rebirth. Only if the ascetic suffers in the profane world, would the Lord bring eternal life in heaven?

Nonetheless, our current times are characterized by a notable lack of certainty and hope for the possibility of a better life. The spread of the secular outlook and atheism seemed to change the cosmology of lay people in regard to natural disasters. Although the dread of a farmer in the tenth century can be analogically compared with a postmodern citizen of a megacity, the point is that our world is much more sorrowful and hazardous. Westerners witness civil wars in the Middle East and parts of Europe as well as higher rates of unemployment and other calamities from the upsurge of a lethal virus to the threat of bio-nuclear terrorist attacks. The postmodern millenarianism sets the pace to panic, pain, and frustration. The rationality and the quest for individual goals engender egoism in citizens. We are living a chronic mourning.

In Kumar's account, modernity prioritized the hermeneutic of the close of a neoliberal utopia. Late modern capitalism has created the annihilation of time and along with it an implicit lack of alternative pathways and emptiness of sense in our day-to-day lives. In this context, Kumar argues that we are in the presence of a devaluated millenarianism that lacks any romantic vision of utopia. Whereas the Christian millenarianism prepares people to live forever through death, the lack of faith in God put our civilization in a troublesome situation because death is contemplated only with fear. Nonetheless, Kumar is convinced the utopia has not completely disappeared in the West but has mutated to other forms such as concern about ecological issues. Whatever the case may be, Christianity paved the way for the advent of a new millennium which can be manipulated by elites to their own benefit. The political manipulation of fear can lead humanity to a real state of conflagration, one fabricated and disseminated by mass media.

Is the apocalypse rational?

One of the intriguing aspects that marks the end of the world seems to be associated with what Eric S. Rabkin dubbed as the introduction of a dark technology. After an erudite review of many myths, Rabkin holds that the world is never fully destroyed; it is reconstructed to purge an original sin committed by greed or curiosity. Doubtless, the man fell, and for that, the earth is redeemed.

In this perspective, the belief in the doomsday or the end of the world comes from original sin. As Eric Rabkin explains, while humans were exiled by God from Eden, they not only were banned from entering this exemplary center but reminded that, someday, the world will be destroyed because of human greed. But the Lord is perfect and he will continue with life, which suggests that the designed destruction calls for a new creation (renovation).

In the written tale of Noah (probably about 850 BCE) God himself recognizes the inevitability of the sins for which He destroyed the World: the Lord said in his heart, I will not again curse the ground any more for man's sake; for the imagination of man's heart is evil from his youth (Gen 8.21). Instead, having purged the world, he urges the remaining people and animals to be fruitful and multiply and premises a new world stability the sign of which is the rainbow, a covenant between me and the earth.

(Rabkin 1983: viii)

When the world is effaced, what disappears is not humankind but only the world as it was preconceived. Society hides some repressed values in such narratives, so by understanding how the world ends, we can comprehend the community behind the apocalyptic myth. This moot point assumes that the apocalyptic contexts prepare peoples towards racism. In fact, the alterity is demonized as an evildoer or simply as a demon who, like Lucifer, wants humankind's obliteration, Rabkin ends. The same ethnocentrism inscribed in the fear instilled by Wells in *The War of the Worlds*, where intruders were described as sexless, blood-sucking and ruthless, remains in the British colonial chronicles about Indians. The question of whether the world has been created (preferably by God) is not limited in the hands of man, though this latter is obliged to administer it (on behalf of God). In almost all narratives, because of greed or imprudence, men offend Gods, for which this world should be purged. However, purgation does not mean total extermination. Life carries on – sublimated – in a new, renascent landscape (Rabkin 1983).

In consonance with this, Gary K. Wolfe introduces the term "Zero Moment" to signal the return to a primitive state of survival where the moral strengths of humankind are at stake. After the world has been destroyed the hierarchal authorities have been suspended as well as all constitutional rights. While a "new aristocracy" arises, Wolfe claims, the impact of technology on human behavior seems to be the key concept lying behind the theory of apocalypse. Civilization was created and evolved thanks to the use of technology; later, the same instrumentalized technology would mine the human soul of corruption and greed. This will be the beginning of the end for the human race. The term, "remaking zero" exhibits the efforts of humans to understand their being in this world and the role of technology in creating a fictional world. As Wolfe reminds us,

The promise inherent in the idea of remaking zero is certainly one of the reasons this genre has survived as long as it has, and in so many guises. On the simple level of narrative action, the prospect of a depopulated world in which humanity is reduced to a more elemental struggle with nature provides a convenient arena for the sort of heroic action that is constrained in the corporate, technological world that we know.

(Wolfe 1983: 4)

Probably the best example that explains how technology affects human freedom is the Matrix Saga. Originally starring Keanu Reeves, Carrie Anne Moss and

Lawrence Fishburne, the saga is situated in a futurist landscape, where human-kind is controlled and enslaved by machines. This great computer imposed a fic-tionalized reality, in which humans believe they live their real lives, while they are actually imprisoned and their bodies exploited as sources of energy for the machines. The Matrix is not only fed by human electrical activity, but there are entire generations born in the Matrix that have never seen the real world. This is exactly the case for Anderson (Reeve), a computer programmer who lacks any interest in his career. When he finds that his life is a complete fake, he is in shock. Anderson knew something was amiss in his life, and this led Trinity (Moss) to introduce him to Morpheus (Fishburne). Morpheus shows Neo two pills, one red and the other blue. Neo is invited to make a decision, as Morpheus puts it. If he takes the red pill, he will face the world as it really is, which would not be a comfortable experience, but if Neo swallows the blue one, he may continue his comfortable life as he does today. Neo takes the red pill, rapidly disappearing and appearing in a new scenario, lying naked in a liquid-filled pod, which is connected to a great machine, the Matrix. Like him, thousands if not millions of bodies lie in other pods connected to this machine, without knowing they are systematically exploited to serve as slaves in a hyper technologized world, where humans will have no place.

This interesting plot tells the history of a rebellion that happened during the twenty-first century, when machines took control of the planet. Humans improved their technologies, introducing artificial intelligence, which created a new type of super entity. In a short and bloody war, humans were pressed to live under the earth. The sun was blocked to deter the advance of the machines and, at the same time, an alternative source of energy was found. Humans were exploited like cattle, the Matrix extracting the necessary bioelectricity to subsist. Prophesized as the One, Neo in one of his missions discovers that he has been chosen to lib-erate humanity from the Matrix's yoke. This revelation does not impede Cypher, a crewmember under the orders of Morpheus, from betraying his fellows, the entire crew, in favor of being reconnected to the Matrix. Cypher – unlike Neo – knows the true life, but it is so harsh that he decides to return to the Wonderland offered by the Matrix. As an anti-hero, Cypher, who colludes with Agent Smith for a stable and comfortable life, disconnects the crew members while they move in the virtual reality, which means a real death, but he fails in the case of Trinity and Morpheus. Far from reaching his objective, Cypher is assassinated by Tank while Morpheus is debriefed by Smith who wants to know the exact location of Zion, the last resisting city. Morpheus finally escapes aided by Neo and Trinity. These are the facts as they happened in the Saga, which captivated the attention of philosophers and social scientists.

In his book, *The Matrix and Philosophy*, William Irwin (2005) reveals that the Matrix evinces the dichotomy between a real life, lived in discontent, or a virtual world, where all needs are met but where the will is annihilated. As Irwin observed, this is not news since Socrates interrogated the need to reach the truth. When Morpheus said to Neo, "Welcome to the desert of the Real", he is empha-sizing Socrates' legacy. While Anderson is rescued from a reality he never lived,

Cypher goes in the opposite direction. He sacrifices his real life for a temporal mitigation of his pain. Both cases suggest two significant points. On one hand, the ongoing quest for happiness or pleasure undermines individual autonomy. On the other hand, the sense of freedom never entails a safe life. From the Cartesian dualism onwards, philosophers erroneously believed that scientific inquiry should be based on rationality, or at best, skepticism. This led us to think that not only might the mind perceive unreal things, but even that the sense of reality might very well remain inexpugnable to the human will. What we perceive probably is unreal, whereas the act of living instills a profound angst, which needs to be alleviated. This is exactly the reason behind Cypher's decision (Irwin 2005). For David Nixon (2005) the belief should be understood as an earlier experience projected in simulacra. Neo was involved in a virtual world, but he believed his life was real. His experiences, elaborated by the Matrix, were real for him, no matter the context. However, this belief never expires, because he validates his trust in Morpheus when he is invited to make a decision. Hence, as Nixon explained, the world of beliefs ends in the decision-making process. In consonance with this, Carolyn Korsmeyer (2005) questions to what extent the real should be unpleasant, or what is worse, if our desire for pleasure-maximization is the factor that gradually ushers us into a climate of alienation.

Whatever the answers to these above questions, Korsmeyer places into discussion that our perception derives from "an illusory program", cementing a long-dormant criticism towards the Enlightenment that emphasized that the gaze was the best instrument to grasp the truth. In this vein, Western civilization adopted an excess of the gaze – presented as a metaphor of understanding – that covered other forms of knowledge. Lastly, the betrayal by Cypher denotes a certain urgency in expunging the past, or the previous traumatic experience, in view of a more pleasurable present.

In sum, the apocalypse can be understood as the rise of reality into the fictional world constructed by humans. However, sooner or later, this fabricated landscape disappears in view of the nature that dominates the earth. Lord Raglan was one of the pioneers in drawing a conceptual model that describes the main character of heroes. They are often persecuted because of their heritage, even on some occasions educated abroad by other families. Their royal lineage puts them between the wall and the deep blue sea often facing major threats and countless obstacles during their mythical trips. Raglan demonstrates amply that the mythical hero exhibits the character of the man who never bowed to the pressure of destiny and God's whim (Raglan 1956). The same archetype can be found in the narratives and discourses revolving around the apocalypse.

As the previous argument shows, Wolfe enumerates some specific features that the end-of-days narratives share, such as the experience of the cataclysm, a journey through the wasteland created by the disaster, the settlement of a new community and the decisive battle to impose human values. Many novels reveal that the central character of the hero remains isolated when the disaster hits. This isolation was in view of an original fault, or simply a moral punishment for an involuntary crime. Almost all heroes are marked (if not cursed) by a sentiment of loneliness which ushers them into depression or madness. The hero should travel far to understand the magnitude of the cataclysm or simply as moral purgation for their personal offenses. This point is vital for a new community to be settled once the hero defeats the forces of evil reminding us one more time why humans are so special. To cut a long story short, let us add that horror movies and the apocalyptic genre are historically intertwined. As we shall see in the next section, the archetype of heroism, as well as the hero's character, prevails in almost all plots.

The world of zombies

Undoubtedly, George Romero was the first film-maker to introduce the figure of zombies in horror movies. In *The Night of the Living Dead* (1968) or *Day of the Dead* (1985), the intriguing double-edge of technology occupies a central role. For some reason, which is not clearly specified in the plot, flesh-eating zombies populate the world, pressing humankind to extermination. The numbers of zombies outnumber humans 400,000 to 1. Dr. Sarah Bowman and Miguel Salazar escape to Fort Myers, Florida where they join others like helicopter pilots John and Bill. However, things go from bad to worse, and they are pressed to make their way towards an army base located in the Everglades. Once there, they are hosted by Captain Rhodes and Dr. Logan, who command a group of scientists and soldiers who undertake surgical experiments with zombies. At this lab, these survivors devote considerable efforts and resources to domesticate (civilize) the cannibalistic zombies. Soon the tension between soldiers and scientists is evident and Rhodes asks the troops to destroy all the experiments in the lab. In an instance, Johnny (the pilot) invites Sarah to drink and relax while Sarah accuses him of being unwilling to subscribe to either side. He replies to her, that the world is exhausted and humans should face the punishment of God because of their greed. Most likely, Johnny bets, the Lord imposed the extermination of humankind to protect the earth because he does not want to see how humans eagerly devour the natural resources. Beyond the extreme suffering, this evokes the restoration of the human race, and of course the planet.

The same can be found in the movie *Resident Evil 2*, where the Umbrella Corporation manipulates a lethal virus that rapidly expands worldwide. The protocols of containment fail and soon the virus is widespread everywhere outside The Hive, the underground laboratory where everything started. Alice and friends struggle against this corporation without knowing that she has been genetically manipulated to be the "perfect warrior". There are many commonalities between the two plots, which merit discussion. On one hand, the use and abuse of technology imply a great risk for humankind. The cannibal "living dead" return to life to eat human flesh while a bunch of survivors struggles to look for another safer place to settle. The forces of evil and good enter in the last of battles, where the values that will reign on the earth are at stake. This mythical conflagration is accompanied by the need to expiate the old human sins. Essential interests are different in the plot of the movie *Contagion* which emulates the latest swine flu (H1N1) outbreak in 2009–2010. After a business trip to China,

Beth Emhoff (Gwyneth Paltrow) returns to Chicago, US to have an adventure with her former lover before going to her family in Minneapolis. She is married to Mitch (Matt Damon) who is not aware of her infidelity. The condition of Beth is suddenly worsened and it finally aggravates her death. Her six-year-old son dies as well. Though Mitch is placed in isolation, he and his daughter seem to be immune. Meanwhile, in Atlanta authorities and scientists of the Department of Homeland Security have a dialogue with Dr. Ellis Cheever (Lawrence Fishburne) to coordinate future efforts to contain and respond to a lethal pandemic. Once again, the sense of fault is rememorized according to the origin and rise of the apocalypse. *Contagion*, unlike *Resident Evil*, signals scientists – not as the evil-doers – but as the heroes who ultimately found and implemented a vaccine that saves millions of lives. The lack of contact between infected and healthy persons as well as the role of mobilities in the configuration of nation-state appear to be some of the topics covered by this film.

Lastly, another interesting movie where the archetype of the end of days can be debated is *Noah*, a film starring Russell Crowe (Noah) and Jennifer Connelly (Naameh). At first glimpse, the plot is developed in a grey post-apocalyptic landscape where the Watchers are punished by God once Semyaza led them to procreate with women. Their offspring, the Nephilim, devoured the cattle and were prone to devastate the earth. God produced a deluge which purged the earth while Semyaza and his fallen angels were castigated. Aronofsky, *Noah*'s director, combines the Chronicles of Enoch book (see on The Watchers) with the classic biblical texts in order to offer a very interesting product. In this plot, the Watchers not only help Noah to struggle against Tubal-Cain but they are the conduit for the Lord's plan to succeed. In fact, God asked Noah to fabricate an Ark, but he is decisively encouraged not to tell God's plan to Cain and his offspring. Noah should select a couple of others to continue with life. The Ark contains not only the germen of life in the next world, prepared by God, but also the future of humanity in the hands of Noah. The problem is that Noah is urged to keep the secret, witnessing how humanity is ruthlessly obliterated through the articulation of a universal flood. This extinction became the first genocide and Noah the architect of a new world where the sins of man are redeemed. The "shameful" Noah falls into depression and drinks all the time until God blesses the family as the new start of a re-born human race. To this point, I have tried to identify the main common arguments in four movies, which were broadcast in different cultural backgrounds and times. Here two interesting points should be outlined. Firstly, as I noted in *The Rise of Thana-Capitalism and Tourism* (Korstanje 2016), Noah was the biblical myth that not only influenced modern audiences but also survived the passing of time. Noah's ark likely revalidates the assumptions that life needs from death. In view of this, Thana-Capitalism rests on the dynamic of creative destruction in the same way Noah is the architect of the first mass cleansing. The drives of death, which today are canalized through dark consumption and dark tourism, are legitimated by this biblical narration. God disposes to exterminate humankind but in secrecy, situating Noah as a privileged witness of his plans. Secondly, what lies beyond the message of this myth is the ongoing competence of the liberal market, which places the rank-and-file worker in egalitarian conditions to others who, like them, will fight to impose their position. Unlike other production systems, where human ties were stable and durable, capitalism gradually erodes the trust liberating the human relations to capital owners. In the days of Thana-Capitalism, citizens rein-force their loyalties in their institutions and their auratic sentiment of supremacy by gazing at how others die. In TV programs, magazines, press, reality shows, novels, movies and other sources of media entertainment (even dark tourism), the Others' death is systematically framed, packaged and distributed as the main commodity to exchange. We are obsessed by consuming suffering while by this rite we are safe at home. This happens for two main reasons. As discussed in earlier sections, the process of secularization not only undermines the influence of religion and the belief in the after-life. Death and the act of dying begin to be conceived of as a sign of inferiority. It is important not to lose sight of the fact that global audiences undergird their sentiment of supremacy while gazing (consuming) at news of disasters, mass death, cleansing and other calamities. They feel part of the "chosen peoples" who are immune to the cataclysm. We toy with the belief this sentiment builds strong narcissism, which unless regulated, may lead towards "chauvinist" reactions. We use the example of *The Hunger Games*, Collins's novel, where the participants (tributes) keep an overrated-image of themselves while fundamentally all will suffer the same end. The rebellion only started when the involved actors opt to cooperate with each other. The same applies to Thana-Capitalism, where the worker is pitted against the worker in a marketplace where a whole portion of yielded wealth is concentrated in a few hands while the rest have a whole life with nothing. Lastly, narcissism, which is stimulated by hyper-consumption and aesthetics, is vital to understand the hege-mony of the ruling elite. The meaning of apocalypse has a double-edged function. While lay-citizens valorize technology as the touchstone of modern society, its negative effects are not discussed. Edward Snowden showed not only how a dem-ocratic government would vulnerate the individual rights to privacy, but also the uses of high tech surveillance when the ideals of democracy are subordinated to fear (Altheide 2014; Skoll 2014; Korstanje & Skoll 2018).

Conclusion

After a further review, this chapter ignited an interesting discussion around the idea of apocalypse or at best the role of technology in slicing humanity from nature. In the terms of the British anthropologist Tim Ingold (2000), one of the concep-tual limitations of "dwelling perspective" as a project is the clear-cut division of humans from the natural environment. Unlike hunters and gatherers who have developed "relational" ties with the sensual world, we are educated to imagine ourselves as administrators of the natural world. In view of this, eco-friendly projects (such as conservation parks) often exclude the presence of humans. The employment of technology denotes artain rationality which, while sorting the environment according to our needs, creates a sentiment of guilt, which is expressed in the end-of-days narratives. Humankind, as an outstanding species, is the only one gifted by the Gods to administer nature, but failed. Moved by greed and speculation, humans governed with energy but turned their backs to God and, for that, they should be heavily punished. As a result of this, the diaspora of survivors appears to be in quest of a new settlement where humans give rise to the new cities, a type of rebirth that alternates the fear of going back to the nightmare of the past and the hope for a better future.

References

Altheide, D. L. (2014). The triumph of fear: Connecting the dots about whistleblowers and surveillance. *International Journal of Cyber Warfare and Terrorism (IJCWT)*, 4(1), 1–7.

Bauman, Z. (2013). *Liquid Fear*. New York, John Wiley & Sons.

Bauman, Z., & Lyon, D. (2013). *Liquid Surveillance: A Conversation*. New York, John Wiley & Sons.

Beck, U. (1992). *Risk Society: Towards a New Modernity* (Vol. 17). London, Sage.

Cohn, N. (1996). "Upon Whom the Ends of the Ages Have Come". In *Apocalypse Theory and the End of the World*, M. Bull (ed.). Oxford, Blackwell, 33–49.

Ellul, J. (1962). The technological order. *Technology and Culture*, 3(4), 394–421.

Ellul, J. (1992). "Technology and Democracy". In *Democracy in a Technological Society*, C. Mitcham (ed.). Springer, Dordrecht, 35–50.

Feenberg, A. (1995). "Subversive Rationalization: Technology, Power and Democracy". In *Technology and the Politics of Knowledge*, A. Feenberg & A. Hannay (eds). Bloomington, Indiana University Press.

Gasparini, S. (2015). "Zombis, Fantasmas, y la representación de la violencia en la narrativa argentina reciente" ["Zombies, ghosts, and the representation of violence in modern Argentina"]. XXVII Jornadas de Investigadores del Instituto de Literatura Hispanoamericana. Facultad de Filosofía y Letras, Universidad de Buenos Aires. Marzo.

Graphs.net. (2013). *Vampires on Social Media*. Accessed March 8, https://graphs.net/vampires-on-social-media.html

Hobbes, T. (2006). *Leviathan*. New York, A&C Black.

Ingold, T. (2000). *The Perception of the Environment: Essays on Livelihood, Dwelling and Skill*. London, Psychology Press.

Irwin, W. (Ed.). (2005). *More Matrix and Philosophy: Revolutions and Reloaded Decoded*. La Salle, Open Court Publishing.

Jameson, F. (1991). *Postmodernism, or the Cultural Logic of Late Capitalism*. Durham, NC, Duke University Press.

Korsmeyer, C. (2005). "Seeing, Believing, Touching, Truth". In *The Matrix and Philosophy*, W. Irwin (ed.). Chicago, IL, Open Courts, 41–52.

Korstanje, M. (2016). *The Rise of Thana-Capitalism and Tourism*. Abingdon, Routledge.

Korstanje, M. E., & Skoll, G. (2018). "Technology and Terror". In *Encyclopedia of Information Science and Technology, Fourth Edition*, Mehdi Khosrow-Pour (ed.). Hershey, IGI Global, 3637–3653.

Kumar, K. (1996). "Apocalypse, Millennium and Utopia Today". In *Apocalypse Theory and the End of the World*, M. Bull (ed.). Oxford, Blackwell, 233–260.

Lazar, S. (2012). How Zombies Became a Social-Media Firebomb. *Entrepreneur*, March 8, www.entrepreneur.com/blog/223075.

McGinn, B. (1996). "The End of the World and the Beginning of Christendom". In *Apocalypse Theory and the End of the World*, M. Bull (ed.). Oxford, Blackwell, 75–108.

Nixon, D. M. (2005). "The Matrix Possibility". In *The Matrix and Philosophy*, W. Irwin (ed.). Chicago, IL, Open Courts, 28–40.

Rabkin, E. (1983). "Introduction: Why Destroy the World". In *The End of the World*, E. Rabkin, M. Greenberg, & J. Olander. Carbondale (eds). Southern Illinois University Press, vii–xv.

Raglan, F. R. S. (1956). *The Hero: A Study in Tradition, Myth and Drama*. New York, Courier Corporation.

Rawls, J. (1999). *The Law of Peoples*. Cambridge, Harvard University Press.

Rousseau, J. J. (1997). *Rousseau: "The Social Contract" and Other Later Political Writings*. Cambridge, Cambridge University Press.

Skoll, G. R. (2014). Stealing consciousness: Using cybernetics for controlling populations. *International Journal of Cyber Warfare and Terrorism (IJCWT)*, 4(1), 27–35.

Skoll, G., & Korstanje, M. (2014). The walking dead and bottom days. *Cultural Anthropology*, 10(1), 11–23.

Thompson, D. (1998). *The End of Time: Faith and Fear in the Shadow of the Millennium*. Lebanon, University Press of New England.

Wolfe, G. K. (1983). "The Remaking of Zero: The Beginning of the End". In *The End of the World*, E. Rabkin, M. Greenberg, & J. Olander. Carbondale (eds). Southern Illinois University Press, 1–19.

Filmography

Contagion. (2011). Steven Soderbergh. 106 minutes. Participant Media, US.

Day of the Dead. (1985). George Romero (dir). 100 minutes. Dead Films, US.

Night of the Living Dead. (1968). George Romero (dir). 96 minutes. Image Ten, US.

Noah. (2014). Darren Aronofsky (dir). 138 minutes. Regency Enterprise, US.

Resident Evil 2. Apocalypse. (2004). Alexander Witt (dir). 94 minutes. Constantin Film, US.

6 Terrorism, tourism and hospitality
Dying in New York City

Introduction

On October 31, 2017, a new terrorist attack shocked the global public in the United States and Rosario city, the second largest city of Argentina. A lone-wolf from Uzbekistan, named Sayfullo Habibullaevic Saipov (29 years old), drove his truck into a group of innocent cyclists killing eight persons and wounding another ten. Among the victims, there were five Argentinians tourist who had originally planned a dream vacation in New York City. The five victims who visited the US in a tight-knit group of friends met their death in Manhattan in a day of terror. They were celebrating the thirtieth anniversary of their high school graduation. Two of these old school buddies were architects, while the others were engineers. Poverty is sometimes used ideologically to explain the reasons for disasters. Of course, the press focused on the frustrations of Saipov as valid explanations of his cruel crime. Meanwhile, journalism revealed that this young terrorist not only had legal residence within the United States, awarded through a lottery, but also was heavily influenced by the videos of ISIS in the media. Is the most democratic nation failing to protect international tourists from the claws of terrorism?

Such a tragic event raises two additional questions: Why are terrorists targeting tourists even using simple transport means? And to what extent does this tragedy enable a sentiment of nationalism in Rosario and Argentina, which can be deciphered through the lens of press and media archetypes?

This chapter attempts to answer the above questions, exploring different disciplines such as geography, anthropology, psychology and sociology. As a multidisciplinary approach, we manifest the desire to continue the discussion begun in the book *Terrorism, Tourism and the End of Hospitality in the West* (Korstanje 2017). To some extent, one of the points of entry in this debate consists in discussing tourism and hospitality as something more than mere leisure industries or forms of entertainment. Tourism, according to Cohen (1979), represents a significant rite of passage, which sublimates the frustrations and inconsistencies of society within an all-encompassing archetype, which alludes to the figure of Lost-Paradise.

Against this backdrop, the first section dissects the nature of tourism as a sacred journey, which – anthropologically speaking – aims to reconstruct the social scaffolding. The second section analyzes the limitations of mobilities and media expansion as two factors that led surely towards a hyper-global world, paving the way for the rise of modern terrorism. In the third section, it is vital to interrogate the nature and evolution of hospitality as a main institution of the West. While hospitality is the main ideological core of Western nation-states, terrorism looks to create political instability by killing tourists worldwide. After all, Saipov was incorporated into American society as a non-Western "Other", who found asylum in the US, invoking the sacred law of hospitality. While he cowardly violated the basis of hospitality, which means the intention of not killing other aliens, the government ignited a debate questioning the lottery for migrants and the current laws of migration in the country. As Korstanje puts it, through the articulation of fear, terrorism harms the essence of hospitality, tightening the borderlands and neglecting the alterity (Korstanje 2017). This explains not only the attacks in New York, perpetrated against innocent cyclists, but also the sentiments and discourses of nationalism long dormant in Argentinean social imaginary.

The importance of tourism

Over the years, popular opinion has taken an unfavorable view of tourism, which is widely labeled as a naïve activity or in the case of some sociologists as a mechanism of alienation (Boorstin 2012). While expeditions, scientific journeys and even the ethnographies of the first ethnologists were certainly based on the needs of discovery, tourist travels, instead, allude to pleasure-maximization goals (Iso-Ahola 1982). Korstanje & Olsen (2011) reviewed almost a dozen plots of different horror movies that saw the light of publicity between 2004 and 2010. They found that the spectatorship is trained to imagine tourists as innocent and vulnerable buddies who needed to escape from the routine of urban populated zones. Because they are unfamiliar with the visited territory, rogues and monsters loom in the dark of the forest. This allegory illustrates one of the incompatibilities of hospitality, where guests are killed or tortured at the hands of psychopathic hosts. It is unfortunate that 9/11 and terrorism have changed not only the ways hospitality is granted but also the plot of horror films. While traveling denotes a certain trust, in view of the fact that hosts and guests are obliged to care for each other, there are a number of ancient myths, ranging from Helen of Troy to Gilgamesh, where the traveling hosts try to kill their guests while they are sleeping after a banquet. Hence hospitality assumes a potential danger for hosts and guests which is mitigated through different elements such as food, drinking and sex (Korstanje & Olsen 2011). Daniel Innerarity holds the thesis that hospitality should be equated to risk. The obsession for a zero-sum society leads gradually to cultural annihilation, while hospitality should be widely granted to all those solicitants who ask for that right. What strangers and risk have in common seems to be the key factor that interpolates to hosting culture widening the opportunities for genuine progress or by closing the borders to what is stable and predictable (Innerarity 2017). Here two assumptions should be made. On one hand, tourism should be discussed beyond the economic-centered paradigm. On the other hand, as a rite of passage, tourism helps in reconstructing social bondage. As Jost Krippendorf puts it, leisure and tourism should be defined as acts of escape that need physical movement and geographical isolation – preferably out of the home – in order for holidaymakers to relax and rest. In view of the fact that the means of production engenders inequalities, asymmetries and cleavages, society elaborates its own forms of control and revitalization, which – like tourism – curb the inter-class conflicts. Through tourism consumption lay-citizens renovate cyclically their loyalties to democracy and the capitalist ethos, or at best, to the mainstream cultural values of the community (Krippendorf 2010). To cut a long story short, in order to work correctly society needs tourism and leisure. Analogically, American anthropologist Dean MacCannell argues that there is a dichotomy between modern and primitive minds. Structuralism contributed notably to social science showing that the sources of authority share no deep differences between modern society and the tribal chieftains. In perspective, structuralism agrees that evolution was the key factor that determined the supremacy of science and modern thinking over other non-Western discourses. At its onset, ancient Europe was in the shadows of superstition and ignorance; this changed when Europe embraced modern reasoning as a criterion of distinction from other cultures. The first ethnographies, conducted in cultures of Melanesia or Australasia, strongly believed that these aboriginal cultures shared resemblance with the life in Europe before the Industrial Revolution. In that way, MacCannell says, the process of secularization eroded the influence of religion in day-to-day life. The figure of totem was a source of power and authority for tribespeople, but it remains empty in modern culture. He toys with the idea that tourism not only fills this gap, left by secularization and the end of religion, but also emulates the role of a totem in modern life (MacCannell 1976, 1984). In this respect, MacCannell confirms the thesis that staged authenticity and consumption play an ideological role as a platform where lay-citizens mediate with their institutions. Equally important, although MacCannell fleshes out an interesting model of tourism, Marxism influences his diagnosis in a pejorative way, and he adds that tourism works as an instrument of domination, which alienates the consumer. As a result of this, the host–guest encounter, far from being genuine, is commoditized by the rules of a market (MacCannell 1973, 1992, 2001).

Last, John Urry (2004) shrewdly pivoted in seeing mobilities as a double-edged sword, opening the doors to expansion and acceleration but at the same time closing the freedom to the hermeneutics of gazing. The car industry provides a good analogy. Though this industry has grown exponentially over recent decades, replacing other classic forms of movement such as horses, trains and ships, the automobile creates certain freedom for drivers while at the same time it coerces them. This means that we can drive faster but only under the ways built for doing so. Mobility is often circumscribed by a set of limited roads which are already fixed to go in the same direction. The advance of modernity brought radical changes not only updating the means of production but also in the patterns of consumption. The modern industrial world successfully monopolized the exploitation of the workforce at the same time as all cultural values were subordinated to a new matrix, which is based on gazing. For Urry and his followers, mobilities and the tourist gaze are inextricably intertwined. This brings two important ideas into the foreground. On one hand, it is unquestionably the fact that the number of journeys has tripled annually worldwide, in which case tourism and hospitality are situated as the main industries worldwide. On the other hand, the travel time has notably shortened in view of faster means of transport. As Urry noted, we start from the premise that in this mobile world, where we live, mobility has been embraced as the main cultural value of capitalism. In this vein, understanding how social cohesion works in an ever-changing society seems to be one of the chief concerns that motivates Urry in his investigation. The analogy of gaze was originally borrowed from Foucault, but in Urry, it takes other connotations. We travel to watch (gaze) and everything which is watched becomes our (symbolic) property. The meaning of the gazed objects corresponds with a much deeper "cultural matrix" that recreates new forms of socialization. Urry contends that tourism – by the exercise of the tourist gaze – expropriates the alterity framing it in specific forms of socialization. Paradoxically, this engenders a complex network of signs and discourses aimed at interpreting what is watched (Urry 2002, 2007; Sheller & Urry 2006). This begs an interesting question: why are terrorists prone to perpetrate their attacks in sites of outdoor recreation and mass consumption?

Terror in a mobile society

Doubtless, terrorism has become a major threat for the West after 9/11 (Dershowitz 2002). For the first time, 9/11 was a test to show the US that its main transport modes, which were the pride of rationality and progress, can be weaponized against civilian targets (Korstanje & Clayton 2012; Saha & Yap 2014; Tzanelli 2016; Seraphin, Butcher & Korstanje 2017). From that moment on, terrorism found in tourism and leisure hot spots a fertile ground to inseminate the kernel of fear (Korstanje 2017).

In the seminal book, *The Political Economy of Terrorism*, Enders & Sandler (2011) offer an interesting economic model to understand terrorism. Although they are not educated in tourism fields, they accept that somehow terrorists focus their attention on international tourist destinations. Beyond the economic losses it supposes, there lies a rational explanation. Moving outside the psychological model, which is aimed at explaining the terrorist's mind, Enders & Sandler provide an economic explanation of the interests of terrorists. At a closer look, the modes of operating in terrorism are not very different than other professions. Like modern citizens, who look to maximize their gains while minimizing the costs, terrorists plan their tactics rationally according to the expectancy of further attention. They invest lowest costs to gain further. Because of its nature – as a leisure activity – tourism provides lower costs and interesting expectancies of further gains. Since the security forces and surveillance technologies retreat at tourist destinations – in order for holidaymakers not to be inhibited – terrorists devote little effort to killing first-world tourists. Secondly and most important, the media effects of terrorist attacks depend not only on the violence and cruelty but also on the vulnerability of the victims. To set an example, when innocent tourists, vulnerable people, youth or children are assassinated society energetically reacts, which means that the status of the victims (their role in the society) seems to be a key factor in the risk perception. The center–periphery dependency is aggravated by the advance of terrorism. Whenever the attacks involve Americans or Europeans the news resonates differently than for third-world tourists. Equally important, the global cities of the North such as New York, London, Brussels or Paris exhibit fertile ground to captivate the attention of the world, whereas other cities of the Global South such as Rio, Buenos Aires or Sydney remain largely anonymous. Gilbert Achcar (2015) dubbed this sentiment as the syndrome of "narcissist commiseration". From his viewpoint, terrorism is something other than a "clash of civilization" (citing Huntington 1993), but a clash of barbarisms, where the cultural radicalization has reached its zenith. The syndrome of narcissist commiseration, Achcar adds, explains the support received by the United States in post 9/11 contexts, while other similar events were mysteriously hidden in the dust of oblivion. For the sake of clarity, the logic of empires is successfully internalized by the periphery simply because the archetype of "exemplary center" blames the victims (the oppressed) instead of the greed of the oppressors. To put this in other terms, citizens of the Global South express their solidarity with Americans narrowing further intimacy with the privileged, sectarian North. Their suffering for the loss and victims of terrorism is masked in a climate of universal suffering, which temporarily and symbolically opens the doors for underdeveloped nations to enter paradise as an equal-other (Achcar 2015). Achcar's contributions are vital to understanding how the material asymmetries that are yielded by the expansion of capitalism are replicated in cultural fields, even in disgrace.

In her project Mega Events as Economies of Imagination, Rodanthi Tzanelli (2016) brings an interesting reflection on the contours of imperialism and hospitality within the Global South. In this respect, mega events connect with historical and local discourses, she says, and terrorism reinforces not only dormant prejudices but also the logic of submission from where Europe historically subordinated peripheral economies. Fear wakes up long-dormant stereotypes which are conducive to the logic of imperialism. The initial stage and evolution of colonialism introduced a form of thinking, rationality, as the best of feasible possibilities, which was externally designed to the non-Western world. The question of whether Europe introduced an aesthetic interpretation of rationality – in the conquest of the world – does not explain its connection with terrorism today. Tzanelli, in this vein, posits some valid answers.

As the previous argument shows, to grasp better Tzanelli's insight two key elements should be mentioned: the artificial economy and the economy of imagination. While the former signals to the needs of normalcy or, at best, familiarity, which denotes home, the latter appeals to the reproduction of allegories which are externally designed but internalized by the hosts. To put this bluntly, the artificial economy conceals not only material asymmetries such as the marginalization of peripheral classes, but also the authority of the nation-state to provide security during the mega event. Equally important, the economy of imagination draws the contours of ideology introducing the exploited classes into the wonderland of consumption. Doubtless, Rio de Janeiro, as the venue for the Olympic Games in 2016, is presented as a good case study to discuss critically the role of tourism and hospitality as sources of a "happier future" for "marginalized *favelados*", who adopt the fabricated and Westernized landscape of a peaceful Brazil. Rio is the archetype of outlaw, representing the image that the white elite conserved of the periphery. Tzanelli writes,

> The state of poverty and the experience of racialization and criminalization remove the act of futurizing from these communities, whereas their displacement in an informal (tourism or otherwise) economy, suggests a sort of phenomenological disappearance from future urban possibilities.
>
> (Tzanelli 2016: 21)

This moot point creates a paradoxical situation simply because the white aristocracy that systematically displaced black and mixed-race people and slaves from the exemplary center in the past, gained further social upward mobility. At this stage, the mobility of few is made possible through the immobility of the rest, but in phenomenological terms, *favelados* are virtually integrated through the orchestration of forced identities. Lastly, Tzanelli interrogates the conceptual incongruences on what Derrida called "conditional hospitality", warning of the risks of multiculturalism. After all, the white guests who were formally invited to Rio 2016 not only interpolate the hosting culture but also revamp the contours of the brilliant future and the dark past. In consequence, those crimes and brutalities that happened during colonial rule are ideologically exchanged towards the formation of a multicultural and global landscape. As Tzanelli puts it, the nations in the Global South adopt tourism not only to legitimate the state of exploitation of some classes but as the precondition towards progress. In consequence, at the same time these economies of imagination advance, the center–periphery dependency enlarges.

Hospitality and terrorism

In this section, we debate the meaning of hospitality avoiding the prejudices and caveats found in the business-related vocabulary. As something that escapes the definition of a mere industry, hospitality can be understood as a "social institution", which regulates the host–guest exchanges (Lynch et al. 2011; Lashley, Lynch & Morrison 2007; Korstanje 2017). One of the most authoritative voices in the analysis of hospitality was Jacques Derrida. Putting his personal imprint, Derrida debates hospitality as the invention of language. At the same time language unites, it repels those who are not fluent or native speakers. In the same way, such a difference may be very well appropriated or not to the extent hospitality can be offered or refused. Beyond the logic of understanding, which is cemented by language, Derrida adds, one of the problems of aliens is they do not speak our language. For this, Derrida equals the law of hospitality within the legal context of the right granted to foreigners. To give hospitality to a solicitant foreigner the State should know who the foreigner is and what he or she wants. According to his development, two types of hospitalities surface: conditioned and absolute. While the latter demands the host opens the home not only to strangers but to unknown travelers, the former is applied only for those who can pay or at least who can exchange something for the lodgings. The absolute hospitality seems to be a utopia as Derrida recognizes, whereas conditioned hospitality calls for a sentiment of reciprocity, which marks politics as the centerpiece (Derrida 2000). Not surprisingly, terrorism endangers the international relations between friendly states in view of the fact that the hosting country should respond to the lack of security for foreign victims. After the terrorist attacks in Bali where more than eighty Australians met their death, Indonesia apologized to the Australian Minister of Foreign Affairs (Howie 2012).

Professor Luke Howie says terrorists do not want a lot of people dying; they want a lot of people watching! In his different approach, he speaks of a new term, known to terrorist experts, which intersects terror with media: witnessing. Taking his original cue from Jean Baudrillard and the spectacle of disasters, Howie maintains that terrorism operates under the horizons of uncertainty, as something other than a mere act of violence – as the specialized literature suggests. Intimidation and coercion would play a leading role in the configuration of a discourse of fear, which is politically tergiversated and manipulated. Emulating the example of celebrities, terrorists need the media attention inasmuch as the tactics are exploiting the vulnerability of global tourists, journalists or businesspeople. The reproduction of terror serves as a disciplinary mechanism that offers an automated idea of the non-Western "Other". Terrorism opens a Pandora's Box, where Muslims are portrayed as dangerous, treacherous and – in some radical voices – a real threat that places the Western lifestyle in jeopardy. Howie adheres to the thesis that terrorism does not exist lest through the lens of media and simulacra. Based on 105 interviews and story-lives, Howie points out that the landscape of 9/11 forged American nationalism henceforth but if further attention is paid, one might realize that terrorism was rechanneled through a previous culture of witnessing which evoked our own vulnerability as city-dwellers. It is important not to lose sight of the rise of Islamophobia related to the impossibility of the modern state to forecast the next attack, but fundamentally, racism was reanimated as a result of the extemporal geographical nature of terrorism. What we watch on TV is happening in a remote city, but the symbolic effects are immediately under our skin as the event is happening here and now.

> The spectacle of terrorism depends on the co-existence of witnesses, images of terrorism, and – in contemporary times – cities. 9/11 happened, it happened on 11 September 2001 in New York City, Washington DC and a field of Pennsylvania. The image, however, is not bound to this temporal and geographic logic. 9/11 was an atemporal event that can be understood in time and space in apparently unlimited coordinates of temporality and spatiality. It resides in the desert of the real of the contemporary city.
>
> (Howie 2011: 60)

By frightening Americans and spectatorship across the globe, media reproduces a "phenomenology of terror" whose effects are enthusiastically accepted for the Global South not only as a form of entertainment but of control. 9/11 started to be a founding event obscuring other similar attacks elsewhere in the world. However, beyond the logic of media entertainment lies a much deeper discourse, which divides the plane between we – the good boys – and they – the monsters who hate us because we are more democratic, multicultural and embrace prosperity in the hands of global capitalism. No less true is that the culture of witnessing aims at producing individual experiences, dotted with a strong ideological discourse, which deserves to be examined (Howie 2011, 2012; Howie & Kelly 2016; Howie & Campbell 2017).

By this token, David Altheide (2017) complements Howie's viewpoint considering that the American culture echoed the terror, which was instilled by terrorism, to form radical politics which now jeopardize the check and balance institutions as well as democracy. The ascendancy of Donald Trump to the Executive Branch reveals not only the power of terror but the decline of hospitality. Altheide is correct when he anticipates that the media reproduce a system of exploitation giving further information to the audience but without any critical scrutiny. There is a strange symbiosis between profits and fear which was conducive to the creation and arrival of ISIS as well as New York's Mogul, Donald Trump. From a preliminary stance, fear would be manipulated by officials to impose economic rules that otherwise would be rejected, but once it arrived, further major changes to democracy are widely accepted. Altheide brilliantly scrutinizes older forms of demonization campaigns in the US, which were orchestrated to avoid the check and balance restrictions. As in other days, 9/11 fostered a climate of patriotism that inevitably led to a clear misunderstanding about the origin of terrorism.

An interesting work, entitled *Hospitality in a Time of Terror*, reminds us that since 9/11 not only has terrorism moved to the center of the stage, but borders have been closed to strangers. Lindsay Anne Balfour (2018) investigates borderlands as spaces of exchange and negotiation, which offers an ethicist reading of Derrida. In a hyper-globalized world, where modern tourism notably contrasts with the urgency of exiles, refugee and asylum-seekers, social scientists should be reevaluated according to a new "aesthetics" post-9/11. Starting from the premise that 9/11 invites us to rethink the contours of hospitality, Balfour adheres to the thesis that 9/11 is called to justify policies that otherwise would be rejected, while a "post-9/11" culture is gradually forged. It poses the challenge of living in a contemporary society which is formed by strangers or "others" while facing the risks posed once these others are welcomed at home. The sense of security, or at best, homeland security, leads us to a climate of paranoia and fear which may very well place hospitality in jeopardy.

Balfour explains that the etymology of the term, hospitality – which is shared with hostility – replicates the philosophical background in a moment of tension between hosts and guests. Revisiting ancient bibliographical sources to trace the origin of the term, in the introductory chapter the author offers a philosophy-based model to understand hospitality as a phenomenological responsibility for the

Other, which puts the self in "radical vulnerability to the other". Certainly, she adds, "hospitality, then, is an invitation to allow the Stranger to remain strange" (Balfour 2018: xviii).

Although her argument rests on Derridean contributions, the days of terror demand a new approach to understanding hospitality not only as a possibility but a decision. Endorsing hospitality to universality Balfour observes that one of the dangers of an "unconditional acceptance" is the dependency the other generates. To put this in other terms, the war on terror – declared by the Bush Adminis-tration – ushered the United States into tightening controls and checkpoints at borders, closing the borders or orchestrating stricter forms of surveillance. Not surprisingly, in movies or novels there is a clear ethical position on the dilemma of "being hospitable" today, reminding us not only that hospitality is in crisis but also in danger. As a result of this – likely the points Derrida glossed over – this impossibility for an open hospitality remains as the cultural centerpiece of hospi-tality itself. The main core of the book aims to remember how the alterity seems to be precise. The chapter toys with the idea that strangers are not enemies but hosts, which structure the cultural self.

In my earlier work, *Terrorism, Tourism and the End of Hospitality in the West* (Korstanje 2017), I theorize on the borders of hospitality and terror. In sharp con-trast with what social imaginary and media preclude, I confront the doctrine of multiculturalism as the only pathway to a safer world. Over time, Europe and the colonial powers used and manipulated mobilities – and of course, hospital-ity – to protect their interests, firstly dispossessing aborigines from their lands, but secondly imposing "the idea of travels" as a criterion of supremacy over the "non-Western Other". Needless to say, the same European paternalism, which facilitated the conditions for the rise of social sciences such as anthropology or sociology, alternated two contrasting discourses respecting the "native". While the romanticism revolving around the allegory of the noble savage cultivated the minds of travelers and even inspired a new genre of literacy, travel-writing, European rule was unilaterally imposed under the auspices of hospitality. Those nations which enthusiastically embraced hospitality as its primary rule not only would benefit from the advance of civilization but paradoxically, the acceleration of globalization mined the efficacy of borderlands as never before. In a united world, the logic of industrialism withered away. With the advance of multiculturalism, nation-states are subject to constant changes. With the passing of years, terrorism silently operated in the margins of the third world but now, echoing Howie, it is situated as a global issue. The chief goals of the terrorist are to produce political instability in Western economies harming the credibility of states. In other words, they found the way to undermine the touchstone of Western civilization (hospitality) from within (Korstanje 2017).

Is terrorism tourism by other means?

To some extent, terrorism and tourism share the same origins. During the nine-teenth century, thousands of European migrants arrived in America in quest of better living conditions. The conditions of work not only were hard, but workers were daily exploited by capital owners. As a result of this, some anarchist voices which infiltrated the mass-migration coordinated efforts to attack politicians and plan terrorist attacks. In particular, the American Presidents, William McKinley, was stabbed by an activist (Leon Czolgosz). Anarchists were rapidly cataloged as "terrorists", jailed, tortured and even killed by the State. There were days of great conflict, tension and violence on American streets. However, gradually other moderate groups made the decision to articulate the organization of worker unions (Joll 1980). Undoubtedly, the anarchists played a leading role in the configuration of the international labor organization and the restructuration of capitalism. Anarcho-syndicalism pressed the government in the quest of improving the miserable conditions of work. The government not only accepted their claims, giving anarchists and their unions the possibility to work fewer hours, but workers received some other benefits which led to modern tourism such as increased wages, holidays and the right to strike (Korstanje 2017). Mentioning the contributions of Foucault (Chapter 4), one might speculate that while the government repressed terrorism like a virus, pushing the anarchist beyond the borders, their main ideological core was internalized by the capitalist system. The fury of the workforce was shrewdly placated by the introduction of leisure and tourism while the disruptive elements were systematically suppressed or exiled. As Foucault (2003) puts it, the disciplinary power aims to dispossess the external threat of its most destructive features, inoculating a nuanced version into the society. He cites the metaphor of the vaccine and the virus. While the former is an inoculated virus which is safe for society to reinforce the natural immunological defenses, the latter represents a serious threat which may place the society in jeopardy. Terrorism (the virus) has been historically disciplined through the articulation of leisure consumption and tourist practices (as a risk) so that the means of production and accumulation proper of capitalism persist. A strike, which was gained as a right of workers, has many commonalities (excepting the violence) with a terrorist attack (Korstanje & Clayton 2012; Korstanje 2017). Firstly, the strike/terrorist attack uses the surprise factor to destabilize the State or capital owners. In doing so, either instrumentalizes the most vulnerable others for achieving their goals. Secondly, strikes are often declared days before a mega event (like the case of the Olympic Games in Brazil), or simply earlier in the summer season (Korstanje, Tzanelli, & Clayton 2014). When the tourists arrive at their destinations they are trapped between a wall and the blue sea, many of them stranded at airports for days, or move of their own accord using their financial resources. The third element consists in the indifference to the Other's suffering. As McCauley & Moskalenko (2008) widely proved, terrorists do not hate their victims; they are not even familiar with them. They developed a negative view of the world which is rechanneled to an abstract hate against abstract objects or discourses such as the end of capitalism, the end of West and so forth. The present case study validates the hypothesis that tourists are ambassadors of their respective nations. Killing foreign tourists not only creates political instability, but places the hosting nation in jeopardy in the eyes of other states. Terrorists (like strikers) are indifferent to the suffering they generate. Of course it is no less true that the strike is legally accepted by the State because it was stripped of the high degree of violence, reminding us of Foucault's main contributions. Within the walls of the city, the strike symbolizes the tension between workers and capital owners while beyond it is dubbed as "terrorism". Fundamentally, tourism and terrorism are two sides of the same coin. For some reason, the process of globalization liberated the repressed risk of terrorism, which was encapsulated in the organization of labor, from its slumber (Korstanje 2015, 2017).

Methodological discussion

Michel Foucault was one of the pioneers in discussing the intersection of language and power. This does not mean other philosophers never explored the theme, but his position provided for the first time a clear-cut explanation of how power molds discourses. He holds the polemic thesis that history and truth not only are social constructs but residual narratives forged and legitimated by the status quo. Paradoxically, while the voice of the privileged elite is embraced as the only valid truth, many other voices are silenced (Foucault 1980, 1982). This suggests that language exhibits not only an ideological nature but also inter-class asymmetries which denote social power. In this respect, Foucault's preliminary insights paved the way for interesting approaches in the analysis of discourse and qualitative methodologies. For some reason, in terrorism-related studies, scholars are not educated, trained or accustomed to using the analysis of discourse as their main method. As L. Stampnitzky (2013) puts it, over past decades terrorism was originally defined as an act of insurgency, in which case, some of the involved parties struggled at a disadvantage against the government. This opened the doors for instrumental definitions, which set the pace to moral judgment in the post-9/11 context. After this event, terrorism was labeled as a moral evil, practiced by evil-doers and maniacs.

As the previous argument indicates, Luke Howie – from Monash University in Australia – alerts us to the limitations of the current applied research in security and terrorism. He toys with the belief that since terrorism is a morally condemnable activity, we are limited to understand the terrorist's mind through the articulation of conventional qualitative research such as story-life or ethnographies. Rather, the position of experts – over recent years – has become speculative and abstract, even TV programs are fraught with "pseudo-experts" more interested in promoting their books than understanding the issue. As Howie eloquently said, the only point the social scientists should follow is how terrorism is discursively portrayed by the media or press, or at best, the effects of terrorism in daily life (Howie 2012). Hence news and the analysis of discourse play a leading role in the configuration of a new paradigm, which helps us to understand terrorism from a novel perspective.

Teun Van Dijk (1988), in his book *News as Discourse*, presents a debate revolving around the ideological discourse behind the objectivity of news. Van Dijk introduces readers to a new theoretical framework for the study of news. Using the analysis of text and discourse, he finds that the production of news – far from being accurate descriptions of fact – is value-laden. As a result, news not only indicates what should or should not be done, but also bespeaks of the values of the ruling classes. The discourse analysis, Ruth Wodak adds, engages a much deeper articulation to differentiate – if not subvert – the relations of power.

> Power is about relations of difference, and particularly about the effects of differences in social structures. The constant unity of language and other social matters ensures that language is entwined in social power in a number of ways: language indexes power expresses power. Power does not derive from language, but language can be used to challenge power.
>
> (Wodak 2007: 11)

As an analytical toolkit, critical discourse analysis (CDA) offers a fertile ground to explore themes, topics which are very hard to grasp. This oscillates from the expression of racism, which is morally and legally repressed, towards the hegemonic discourse beyond terrorism. Most likely, CDA is often misinterpreted as a method instead of a technique to process information. As a perspective, it is enrooted in the critical analysis of discourse (critical discourse studies) which confronts multivariable problems. CDA rests on the need to discover the veil of power, as masked in the logic of discourses. After all, those groups which devote considerable efforts in monopolizing the production of meaning as well as the circles of knowledge are likely to remain unquestioned, exerting in this way a most influential control over the rest. CDA helps to decode the lines of domination-submission, exploring the role of those leaders pivotal to the production of ideology within organizations (Van Dijk 2015). Van Dijk goes on to say that,

> If controlling the contexts and structures of text and talk is a first major form of the exercise of power, controlling people's minds through such discourse is an indirect but fundamental way to reproduce dominance and hegemony. Indeed, discourse control usually aims at controlling the intentions, plans, knowledge, opinions, attitudes, and ideologies – as well as their consequent actions – of recipients. A socio-cognitive approach in CDA thus examines social structures of power through the analysis of the relations between discourse and cognition. Cognition is the necessary interface that links discourse as language use and social interaction with social situations and social structures.
>
> (Van Dijk 2015: 472)

One of the authoritative voices in CD analysis, Siegfried Jager (2007), exerts a radical criticism of Michel Foucault because according to his diagnosis, the French philosophers conceptually dissociated language from practice. In any case, discourses vary over time, but with them, the object is being changed. Since Foucault defines discourses as fictionalities or disconnected from reality, Jager observes, he overlooks that while discourses are changing, objects do not change their constructed meaning but lose their "previous identity" (Jager 2007: 43). In this vein, similar events – as in the case of nuclear accidents – can be conceptualized in different ways according to their "discursive context". In a nutshell, he goes on to say,

> All events have discursive roots, in other words, they can be traced back to discursive constellations whose materializations they represent. However, only those events can be seen as discursive events which are especially emphasized politically, that is as a general rule by the media, and as such events, they influence the direction and quality of the discourse.
>
> (Jager 2007: 48)

Some interesting outcomes to the application of CDA in terrorism issues may be found in the book *Violencia de Texto, Violencia de Contexto* (*Violence of Text, Violence of Context*) authored by Chilean historian Freddy Timmermann. He dissects the official documents issued during Pinochet's government to fight against the insurgency. Through the qualitative and critical analysis of these texts, adjoined to the reading of survivors' or prisoners' testimonies, Timmermann concludes that the climate of violence which in Chile allowed the rise of Augusto Pinochet to power was fed through the orchestration of a long-dormant fear against communism (red-scare), which dismantled not only the worker unions' opposition but also changed forever the ways lay-citizens lived politics. Timmermann analyzes two books written by Rolando Carrasco, who was a "disappeared" dissident tortured by the government in the 1970s. Through the reconstruction of Carrasco's text, he successfully draws the borders of the socio-economic background of Chile, which was the precondition for the advance of terrorism and political repression (Timmermann 2008).

Terror in New York: from the spectacle to the oblivion

Though Buenos Aires suffered two important terrorist attacks in 1992 (in the Israeli Embassy) and in 1994 (at AMIA, Asociacion Mutual Israelita Argentina) resulting in 114 deaths, these events did not receive the same emotional treatment in the media as the recent attack in New York, where five Argentinean tourists died. The current president Mauricio Macri not only showed formal interest in this event, visiting a memorial tribute at the site as well as speaking with Donald Trump about future coordinated efforts in the struggle against terrorism, but also valorized friendship as the main cultural value of Argentinians. Metaphorically speaking, after decades of isolation from the world, Argentina entered this dangerous world with a baptism of fire.

To clarify, let's explain that Argentina is geographically situated in South America sharing borders with Chile, Uruguay, Paraguay, Bolivia and Brazil. The official language is Spanish and the country has an area of 2,780,400 km^2. As the eighth largest country in the world, Argentina is subdivided into twenty-three provinces.

Over recent years, the growth of inflation associated with other economic problems which were accelerated after the stock market crisis in 2008, resulted in a
climate of speculation that led towards an irreversible depreciation in consumer
purchasing power. Thousands of tourists, henceforth, visited other destinations
which were historically reserved by the aristocracy or higher classes such as New
York, Madrid, Paris or even the paradisiacal Miami. As stated, New York gradually replaced the classic destinations such as Santiago or Madrid as the first destination of Argentinian tourists in the past five years. According to the United States
National Tourism Office Travel and Tourism Indicators, almost 60% of Argentinian citizens who entered the country visited New York. Per capita, tourists spend
approximately USD 2,521 with an average stay of nine days.

It is unfortunate that on October 31, 2017, a young man of 29 years, Sayfullo
Saipov drove a truck for nearly a mile through lower Manhattan, striking pedes-
trians and cyclists. Among the victims were five Argentinian tourists. The five victims were from Rosario city, Argentina: Hernán Diego Mendoza, Diego Enrique
Angelini, Alejandro Damián Pagnucco, Ariel Erlij and Hernán Ferrucci. These
tourists were part of a group of ten, who visited New York with school friends
to celebrate the thirtieth anniversary of their graduation. The attacker was caught
alive and imprisoned to await his trial. Not only does nobody think they will die
on holiday, but this sacred journey – as discussed in an earlier section – emanates
from the source of authority of contemporary society. Hence this tragedy cast
a shadow over Argentina and especially Rosarinos (Rosario dwellers); the war
on terrorism became a cause and a priority for Mauricio Macri's presidency. In
the lines to follow, we will sketch the most significant discursive utterances and
columns from the Argentinian newspapers and magazines to draw a conceptual
model to understand the *argentinidad* ("Argentinianness"), as well as reciprocity,
that activates the social imaginary about New York City as an exemplary center.

On November 1, 2017, *Clarin* (one of the most read newspapers) released a
column entitled "Cinco Argentines Muertos pour un nuevo atentado terrorists
en Nueva York" ("Five Argentinians assassinated after a new terrorist attack in
NY"). On closer inspection, the article emphasizes the declarations of Mauricio
Macri who expressed his condolences to the families of the victims, "We, the
Argentinians, are profoundly shocked by these tragic casualties in NY. I stay to
the disposition of families who had lost their relatives".

The lines above evince the construction of a symbolic bridge between Argentina and the US, which are emotionally united by a tragedy. Bill de Blasio, Mayor
of New York, at the same time confirmed the hypothesis of a terrorist attack
expressing his solidarity with the victims. However, de Blasio attempted to calm
popular opinion clarifying that this was the act of a lone-wolf, which suggests
no others cells are operating in the city. The figure of the lone-wolf acts as a
form of domesticating the future, giving information to control the high levels of
anxiety in the population. As discussed, when a terrorist attack hits, people are
alarmed simply because the same may very well happen anytime and anywhere.
A second point in the analysis seems to be the origin and ethnicity of the young
terrorist. He was born in Uzbekistan and arrived in the US thanks to a lottery that

permitted his legal entrance. This was seriously questioned by President Donald Trump who energetically asked the migration office to constrain the current visa lottery program. In consonance with Korstanje (2017) and Altheide (2017), terrorism encouraged radicalized responses which are oriented to tighten the controls at the borders. This seriously affects not only tourism but also other industries. To this end, Macri in Buenos Aires declared openly that "there is no place to timid responses or gray zones at the time to fight against terrorism, we have to engage in the war on terror from today on". The speech was at the opening of a GAFI meeting, Grupo de Accion financiers, at the Hilton Hotel.

Against the previous backdrop, the emotional factor was evoked in the successive days while journalism emphasized the individual stories of the victims and their biographies as well as their long friendship.

The survivors dressed in white T-shirts in commemoration of their friends who were assassinated in the attack. At a first glimpse, here the essence of hospitality appeals to the emotionality of what being Argentinian means. The Argentinian ("Argentinianness") is culturally constructed according to the values of loyalty and friendship, which is cultivated from the cradle. This mythological narrative was not only emulated by the media and press but President Macri as well. In contrast to Americans who are shown as cold and afraid of strangers, Argentinians are hospitable people who not only love strangers but also keep an "open mind". Such a "discourse" emphasizes the need to keep traveling the world and not to surrender to the panic terrorists want to impose. Macri received a phone call from Trump and days later traveled to New York to bolster business meetings with some important American businesspeople. Once there, Macri headed a ceremony in tribute to the victims at the site where the bodies were found. Like the site of dark consumption or dark tourism, the site exhibited the Argentinian football team, accompanied with flowers and candles.

Infobae, an Argentinian news website, traced the biography of the terrorist back to his ideological dependency on ISIS, as well as the videos he watched which supposedly inspired the massacre. The media, once again, played a leading role constructing the attachment between ISIS and its future candidates. Saipov had asked for an ISIS flag to decorate his room at the hospital but this was denied by the authorities, Infobae reported. This begs an interesting philosophical question, which merits further reflections. On one hand, the anniversary of these ten friends – which was a rite of passage orchestrated to foster in-group cohesion – was frustrated by the failure of the most powerful state to bring hospitality to stranger tourists, but at the same time, the murderer was hosted to receive health treatment. The metaphorical hotel (commercial hospitality) and hospital (absolute hospitality) converge with the lack of any sentiment of remorse by Saipov, the monster who transformed hospitality into the opposite, hostility. The act of traveling abroad – abandoning home – demands greater levels of uncertainty, because the traveler is unaware of the visited destination or simply because less is known about the host's intentions. Saipov received the same hospitality he rejected for the five victims. This evinces the paradoxical situation which activates terrorism, widely studied by Korstanje (2017). At the same time, tourism and globalization connect an ever-changing world – through the orchestration of different automated and faster means of transport, robots and other devices, media consumption and so forth – but such destinations attract the attention of terrorism. The same attractiveness, citing Howie (2012), these real exemplars of paradise show, increases the probability of an attack. Here is when the paradox revolving around the nature of hospitality surfaces, oddly by limiting the ethical response of economic power to the actual migratory crisis or even facilitating Islamophobia in white culture. While the case of Alan Kurdi, the Syrian boy found dead on the coast of Turkey, created huge sympathy worldwide for the situation of refugees and asylum seekers, terrorism mined the common trust, demonizing the non-Western Other as an enemy of the State. This chapter intended to reflect the ebbs and flows of hospitality as well as the contours of reciprocity among states.

Conclusion

As a rite of passage, tourism revitalizes the psychological frustrations that occur during working time. During holidays the rules are temporally subverted in order for citizens to accept unilaterally the authority of nation-state, cultivating not only the passion for leisure but also the importance of labor. If this rite is not successfully achieved, the legitimacy of officialdom may be seriously hurt. In this vein, terrorism operates under the dark side of anonymity, while a great majority of terrorists are legal residents or native-born in the country they hate. Carefully, and after evaluating the costs and benefits of killing tourists, terrorists target leisure hot spots and paradisiacal destinations not only to affect the economy of the hosting nation but also to erode the symbolic totem of Western civilization: hospitality. In the years to come, this point would be taken seriously into consideration by experts and policy-makers to design safer destinations and products. This chapter reminds us that the essence of evilness consists in the lack of ethics that grants hospitality, in the same way that God will protect us in the afterlife.

References

Achcar, G. (2015). *Clash of Barbarisms: The Making of the New World Disorder*. Abingdon, Routledge.

Altheide, D. (2017). *Terrorism and the Politics of Fear*. New York, Rowman & Littlefield.

Balfour, L. A. (2018). *Hospitality in a Time of Terror*. Lanham, Bucknell University Press.

Boorstin, D. J. (2012). *The Image: A Guide to Pseudo-Events in America*. New York, Vintage.

Cohen, E. (1979). A phenomenology of tourist experiences. *Sociology*, *13*(2), 179–201.

Derrida, J. (2000). *Of Hospitality: Anne Dufourmantelle Invites Jacques Derrida to Respond*, trans. Rachel Bowlby. Stanford, Stanford University Press.

Dershowitz, A. M. (2002). *Why Terrorism Works: Understanding the Threat, Responding to the Challenge*. Ithaca, New York, Yale University Press.

Enders, W., & Sandler, T. (2011). *The Political Economy of Terrorism*. Cambridge, Cambridge University Press.

Foucault, M. (1980). *Power/Knowledge*, Colin Gordon (ed.). New York, Pantheon.

Foucault, M. (1982). The subject and power. *Critical Inquiry*, *8*(4), 777–795.

Foucault, M. (2003). *"Society Must Be Defended": Lectures at the Collège de France, 1975–1976* (Vol. 1). New York, Macmillan.

Howie, L. (2011). *Terror on the Screen: Witnesses and the Re-animation of 9/11 as Image-event, Popular Culture and Pornography*. Washington, DC, New Academia Publishing, LLC.

Howie, L. (2012). *Witnesses to Terror: Understanding the Meanings and Consequences of Terrorism*. New York, Springer.

Howie, L., & Campbell, P. (2017). *Crisis and Terror in the Age of Anxiety: 9/11, the Global Financial Crisis and ISIS*. New York, Springer.

Howie, L., & Kelly, P. (2016). Sociologies of terrorism: Holographic metaphors for qualitative research. *Journal of Sociology*, *52*(2), 418–432.

Huntington, S. P. (1993). The clash of civilizations? *Foreign Affairs*, *1*, 22–49.

Innerarity, D. (2017). *Ethics of Hospitality*. Abingdon, Routledge.

Iso-Ahola, S. E. (1982). Toward a social psychological theory of tourism motivation: A rejoinder. *Annals of Tourism Research*, *9*(2), 256–262.

Jager, S. (2007). "Discourse and Knowledge: Theoretical and Methodological Aspects of a Critical Discourse and Dispositive Analysis". In *Methods of Critical Discourse Analysis*, R. Wodak & M. Meyer (eds). London, Sage, 32–62.

Joll, J. (1980). *The Anarchists*. Cambridge, MA, Harvard University Press.

Korstanje, M. E. (2015). The spirit of terrorism: Tourism, unionization and terrorism. *Pasos*, *13*(1), 239.

Korstanje, M. E. (2017). *Terrorism, Tourism and the End of Hospitality in the West*. New York, Palgrave Macmillan.

Korstanje, M. E., & Clayton, A. (2012). Tourism and terrorism: Conflicts and commonalities. *Worldwide Hospitality and Tourism Themes*, *4*(1), 8–25.

Korstanje, M. E., & Olsen, D. H. (2011). The discourse of risk in horror movies post 9/11: hospitality and hostility in perspective. *International Journal of Tourism Anthropology*, *1*(3–4), 304–317.

Korstanje, M. E., Tzanelli, R., & Clayton, A. (2014). Brazilian World Cup 2014: Terrorism, tourism, and social conflict. *Event Management*, *18*(4), 487–491.

Krippendorf, J. (2010). *Holiday Makers*. London, Taylor & Francis.

Lashley, C., Lynch, P., & Morrison, A. J. (Eds). (2007). *Hospitality: A Social Lens*. Oxford, Elsevier.

Lynch, P., Molz, J. G., Mcintosh, A., Lugosi, P., & Lashley, C. (2011). Theorizing hospitality. *Hospitality & Society*, *1*(1), 3–24.

MacCannell, D. (1973). Staged authenticity: Arrangements of social space in tourist settings. *American Journal of Sociology*, *79*(3), 589–603.

MacCannell, D. (1976). *The Tourist: A New Theory of the Leisure Class*. Berkeley, University of California Press.

MacCannell, D. (1984). Reconstructed ethnicity tourism and cultural identity in third world communities. *Annals of Tourism Research*, *11*(3), 375–391.

MacCannell, D. (1992). *Empty Meeting Grounds: The Tourist Papers*. London, Psychology Press.

MacCannell, D. (2001). Tourist agency. *Tourist Studies*, *1*(1), 23–37.

McCauley, C., & Moskalenko, S. (2008). Mechanisms of political radicalization: Pathways toward terrorism. *Terrorism and Political Violence*, *20*(3), 415–433.

Saha, S., & Yap, G. (2014). The moderation effects of political instability and terrorism on tourism development: A cross-country panel analysis. *Journal of Travel Research*, *53*(4), 509–521.

Séraphin, H., Butcher, J., & Korstanje, M. (2017). Challenging the negative images of Haiti at a pre-visit stage using visual online learning materials. *Journal of Policy Research in Tourism, Leisure and Events*, *9*(2), 169–181.

Sheller, M., & Urry, J. (2006). The new mobilities paradigm. *Environment and Planning A*, *38*(2), 207–226.

Stampnitzky, L. (2013). *Disciplining Terror: How Experts Invented "Terrorism"*. Cambridge, Cambridge University Press.

Timmermann, F. (2008). *Violencia de texto, violencia de contexto: historiografía y literatura testimonial. Chile, 1973*. [*Violence of Text, Violence of Context, Historiography and Testimonial Literature in Chile 1973*.] Santiago, Ediciones Universidad Católica Silva Henriquez.

Tzanelli, R. (2016). *Mega Events as Economies of Imagination: Creating Atmospheres for Rio 2016 and Tokyo 2020*. Abingdon, Routledge.

Urry, J. (2002). Mobility and proximity. *Sociology*, *36*(2), 255–274.

Urry, J. (2004). The "system" of automobility. *Theory, Culture & Society*, *21*(4–5), 25–39.

Urry, J. (2007). *Mobilities*. Cambridge, Polity Press.

Van Dijk, T. (1988). *News as Discourse*. Hillsdale, Lawrence Erlbaum Associates.

Van Dijk, T. A. (2015). "Critical Discourse Studies: A Sociocognitive Approach". In *The Handbook of Discourse Analysis*, D. Tannen, H. Hamilton, & D. Schiffin (eds). Oxford, Wiley Blackwell, 63–74.

Wodak, R. (2007). "What CDA is About: A Summary of its History, Important Concepts and its Development". In *Methods of Critical Discourse Analysis*, R. Wodak & M. Meyer (eds). London, Sage, 1–13.

7 Democracy and its faces

The problem of Islamophobia

Introduction

From Gordon Allport onwards, psychologists, anthropologists and sociologists have adopted a coherent definition of prejudice and racism. In his seminal book, *The Nature of Prejudice*, Allport (1979) introduces a much deeper conceptual debate with concrete experiments and case studies. Allport is convinced that individuals use stereotypes to sort the world. As the external world is chaotic and complex, it should be classified through the articulation of some cognitive constructs, which may be the origins of prejudice and racism (Allport 1979). Allport helps to reveal the emotional nature of discrimination while illuminating colleagues and pundits to date (Tajfel 1969, 1981). Some scholars emphasized the authoritarian character of racists, whereas others signaled the social deprivations and psychological frustrations as the key factors that explain prejudice. Over recent years, some cultural theorists such as Teun Van Dijk lamented that prejudice is far from disappearing, but has mutated to a new (subtler) cultural form. The concept of race, as imagined by Darwinists in the nineteenth and twentieth centuries, was replaced by cultural superiority (Van Dijk 1992, 1993, 2000). That way, some classes or ethnic groups are considered inferior in view of their lack of cultural maturation compared with the ruling classes. Norbert Elias showed that there is a manifest tension between established families and new ones in his study of "Winston Parva" in the UK. Although neither faction evinced a clear distinction respecting class, race or status, the privileged position of already settled families created a symbolic circle of discrimination for outsiders and new residents (Elias & Scotson 1994). All this suggests that even if prejudice seems to be inherent in the human mind to protect some symbolic barriers, it can be culturally encouraged or discouraged.

Today, Islamophobia has displaced classic racism in the concept of race but conserved all the old semantic structure, which means both racisms share similar conditions and discourses, though a subtle change relates to the epoch (Saeed 2005). For some reasons, white social scientists pay little attention to Islamophobia, while the problem is widely analyzed by Muslim-related scholars. This chapter reflects on the scourge of this new racism calling attention to the philosophical dichotomies of a closed hospitality and the refugee crisis in the Middle East. Recently, Donald Trump promoted the Travel Ban, a set of legal restrictions for travelers and tourists coming from nations suspected of terrorism. Though it was declared unconstitutional by the Supreme Court, some interesting questions arise. Is the US a nation which is ceding to totalitarian expressions? Are Americans racists? Is Islamophobia a type of new racism or simply the fear terrorism has manipulated? And why is the West closing itself to the "Other"?

Elizabeth Poole, one of the most authoritative voices in Orientalism, held that Islamophobia represents a new racism which through the floating signifier was hosted in European ethnocentrism. The term Islamophobia denotes something other than a simple fear; it assumes that there is an increasing anti-Muslim sentiment justified by the concerns some Europeans feel regarding terrorism and the Islamic State. The mass media replicates the cultural hegemony of the ruling class, labeling and closely controlling some minorities as well as reinforcing old prejudices, Poole (2002) overtly says.

With this in mind, this chapter confronts Foucault and his medical gaze in understanding not only the nature of Islamophobia but the future of Muslims in Europe. Our thesis is that the medical gaze looks for the pathology, which means what is being corrupted, ill within the organism. For example, take the case of cancer, which includes a set of diseases involving abnormal cell growth potentially invading other organs of the body. In an earlier chapter, we were reminded that Foucault (2003) has outlined the roots of disciplinary power, as originally aimed at domesticating external dangers. However, he glossed over what happens when the society fails in keeping away the looming threats. Once one of the organs is infected, the impossibility of the medical gaze, adjoined to the impossibilities of the medical gaze to elaborate a cure, leads to the extirpation of the organ. The same very well applies to the Muslim community, which runs serious risks of being humiliated, segregated and even deprived of some rights if terrorism cannot be contained (Korstanje 2017). Jacques Derrida, in this direction, defines terrorism as an autoimmune disease where the same antibodies attack the immune system of the society (Borradori 2013). As demonstrated in the earlier chapters, terrorism rests on the instrumental rationality coined by the West, connoting a means-and-ends logic. Terrorists do not want a lot of people dying; they rather want a lot of people watching (Howie 2012, 2015; Howie & Campbell 2017). This poses serious challenges about how the communicative process is handled in democratic societies. An excess of censorship as well as an excess of publicity may seriously harm the reputation of government. M. Eid (2014) uses the word *Terroredia* to symbolize the conjunction of terror + media. While the media corporations need further investors to survive, terrorists are seeking lower cost sources to make their message public. In a strange symbiosis, the media is obsessed by covering news containing terrorist attacks at the same time as terrorists have found in the media a fertile ground for achieving their goals. Paradoxically, this recreates the contours of an ideological discourse, which has unfortunately ended in the "subordination" of the Muslim community in the Global North.

Preliminary debate

Miguel Bandeira Jerónimo and Jose P. Monteiro (2017) examine the interplay between imperialism and internationalization. They call attention to the need to cross the historical barriers to reach an all-encompassing model of the problem. Their erudite edited book, titled *Internationalism, Imperialism and the Formation of the Contemporary World*, is situated as an attempt to discuss imperialism from diverse angles, but more importantly places the term into its specific socio-cultural background. Internationalism should be seen neither as a simple sum of variables nor a sum of scattered parts. In fact, as the editors said, the theory of imperialism has been dominated by the national spirit for many years, which obscures much of the produced knowledge in the academic circles. Political scientists have been heavily influenced by the archetype of nation-hood, which has led them to distorted and biased conclusions about Empires.

> Overcoming the boundaries of nation-state requires the study and rethinking of the flows and interactions that have crossed them, but also the mental universes and the political and ideological frameworks of those who have promoted the creation of programmes, groups, or communities whose world of connection and scope of action did not respond to strict or predominantly national criteria.
>
> (Jerónimo & Monteiro 2017: 9)

As the above quotation indicates, empires evolved according to a complex framework which disposes of the colonial minds, bodies and goods territorialized and re-territorialized in symbolic and material terms. Empires expand themselves through a social discourse, which is aimed at subordinating the conquered cultures to the imperial matrix. As discussed, internationalism, as well as imperialism, takes different shapes according to the cultural conditions and periods.

With this background in mind, A. O'Malley (2017) dissects the relations of the United Nations as a supra-corporation, which rules or dictates the course of actions among others nations and Anglo-America. Based on the case of the Congo mission promoted by the UN, O'Malley writes that the country entered an unprecedented crisis after independence from Belgium on June 30, 1960. The outbreaks of violence pushed Belgium's king to send troops to protect the Europeans living there. The Prime Minister, Patrice Lumumba, called on the UN to intervene to prevent the invasion of Belgium, which the Congolese Government defined as a unilateral act of aggression. On one hand, the UN authorized the delivery of a peace commission whose aims were the restoration of peace. Understood as an example of the decolonization process, the two remaining superpowers (Soviet Union and the US) articulated some tactics to protect their interests in the region. The UN finally took direct action and the crisis in Congo showed the limitations for peripheral African nations to keep some autonomy in the context of an emergency or economic crisis. As O'Malley reflects, the treatment of humanitarian emergencies is not equal, but accords to the status or the position of the nation on the UN Security Council. Paradoxically, while European powers and the UN delivered missions to pacify Congo, this cast serious doubts on Congo's autonomy. In the name of humanitarian interventions, some violations of sovereignty can occur. On the other hand, in the name of democracy, like the case of Adolf Hitler who adduced humanitarian issues while he ordered the invasion of Czechoslovakia, there are no clear boundaries that regulate with accuracy the limits of human rights. The invention of patriotism, which is legitimated by a biased view of history, may very well be conducive for dictatorships to flourish. This leads to the question: how human are human rights?

Doubtless, there is no case that better reflects the contradictions of democracy than Catalunya independence. While Carles Puigdemont, the former Catalan president, invokes their right to be independent from Spain, Mariano Rajoy, Spain's Prime Minister, appeals to the respect for law to keep separatists under control. Therefore both sides allude to democracy to protect or validate their own interests, to the extent that the majority of the Spanish people are pushed into an unnecessary confrontation. The sense of state is not only based on reciprocity, as was discussed in Chapter 1; but it is also in an idea of unity, which is culturally and symbolically constructed (Rajan 2011). The stock market crisis accelerated a state of decomposition, which may very well lead towards a new feudalism or climate of separatism as never before.

The clash of cultures

One of the most polemic approaches that theorized about Muslim civilization, as well as the open risk of a clash between West and East was *The Clash of Civilizations*, a thesis originally developed by American political scientist Samuel Huntington. Though he proposes a polemic theory revolving around terrorism, it is no less true that his thesis resonated in the Western social imaginary just after 9/11 as never before. He toys with the idea that civilizations are historically constituted to clash. This would-be natural trend was accelerated after 9/11 and after the Bush administration made the decision on the US-led invasions of Afghanistan and Iraq. As Huntington overtly lamented, part of the resentment against the West is previously determined by the lack of progress, prosperity and liberty as well as the state of political instability of the "failed states". With the benefit of hindsight, he goes on to say,

> In this new world the most pervasive, important and dangerous conflicts will not be between social classes, rich and poor, or other economically defined groups, but between people belonging to different cultural identities. Tribal wars and ethnic conflicts will occur within civilizations. Violence between states and groups from different civilizations, however, carries with it the potential for escalation as other states and groups from these civilizations rally to the support of their kin countries.

(Huntington 1997: 28)

Huntington's legacy is a fertile ground to accept the scatological idea that 9/11 inaugurated a new epoch, leading Western civilization into a difficult position. Through his insight, civilizations are centered on constituent features, which are at stake with other cultural values. Inevitably, from its inception, any civilization is created and destined to struggle with others to survive. Democratic societies such as in the US and Europe consolidated a state of prosperity, not only by their success in updating their productive systems, but also by the political stability they cultivated over the years. As Huntington puts it, the established democracies successfully prevented large-scale conflict or civil wars. The fact is that, with a closer look, this does not happen in the Middle East or Latin America, where the nation-states systematically failed to promote a sustainable political system. The fall of the Soviet Union reminds us that the world has been fractured into seven civilizations: Latin American, African, Islamic, Sinic, Hindu, Orthodox, Buddhist, Japanese and of course the West shaped by USA, Australia and Western Europe.

In his argument Huntington acknowledges that the success or failure of democracy as a supreme value depends to a greater or lesser degree on the cultural structure of the country. The spirit of democracy in the Middle East was unfeasible because of the action of Islam as the main religion. Taking his cue from Francis Fukuyama, Huntington strongly believes that after the "end of history" the prosperous democratic ideals would extend worldwide. This does not mean that totalitarian regimes would completely disappear, but sooner or later, the democratic nations – which are gradually being pitted against autocratic states – will prevail. Above all, the ideals of a Kantian perpetual peace, which was rooted in the Enlightenment, finds in religion a main obstacle to overcome.

> The idea of civilization was developed by eighteen-century French thinkers as the opposite of the concept of barbarism. Civilized society differed from primitive society because it was settled, urban and literate. To be civilized was good, to be uncivilized was bad. The concept of civilization provided a standard by which to judge societies, and during the nineteenth century, Europeans devoted much intellectual, diplomatic and political energy to elaborating the criteria by which non-European societies might be judged sufficiently civilized to be accepted as members of the European-dominated international system.
>
> (Huntington 1997: 41)

Any civilization meets a cultural archetype that is orchestrated through a process of territorialization, in which case, norms, traditions, technology, values and mythology collide. In this respect, Huntington rejects the classic definition of race as the articulator of white supremacy, but he observes, the prosperity of English speaking countries was legitimated in the maturation of an economic system, which is based on the needs of promoting egalitarianism and freedom as mainstream cultural values (Huntington 1993, 1997). Hence, democracy appears to be the cornerstone of American supremacy over other (less democratic) cultures such as Latin America, Africa or Asia. This case sounds particularly interesting because despite sharing the same (European) roots, Latin Americans were trapped amidst economic stagnation and the successive alternations of democratic governments and military coups.

> Latin America, however, has a distinct identity which differentiates it from the West. Although an offspring of European Civilization, Latin America has evolved along every different path from Europe and North America. It has had a corporatist, authoritarian culture, which Europe had to a much lesser degree and North America not at all.
>
> (Huntington 1997: 46)

Here I found some of the contradictions in Huntington's argument. At first glimpse, Huntington – beyond his eloquence – never provides concrete indicators or evidence that proves democracy is conducive to economic prosperity. Growing nations today, whose economic position is better than the US, have accumulated various denunciations for the violation of human rights. Equally important, Huntington glosses over the fact that democracies can be as totalitarian as monarchies and vice versa. Huntington's concerns regarding the invasion of a foreign language in the US are unquestionable. Whereas the elite exerts influence on the populace with respect to certain fashionable tendencies, a foreign language marks the difference between aristocracies and the rest of the population. To what extent should civilization clash? Huntington is not offering a convincing answer nor a clear-cut investigation that leads towards conclusive evidence. The process of modernization in the West, historians agree, should be explained by means of interaction with other regions, not as a constituent feature in itself. Civilizations are formed according to their contact with other civilizations. On one hand, the so-called superiority of the West is given in terms of the material conditions of production.

On the other hand, contact and fluid dialogue with other nations engendered, following Huntington, a spirit of free trade, new techniques in innovation and the necessary technological breakthrough to mark a difference with other cultures. Those agrarian societies that forged a centralized authority were impermeable to trade and democracy, basing their politics on the landowners and their rights of land tenure. Industrial-based societies embraced democracy as their primary form of government. Over recent years, the West has witnessed how the expansion of modernity has consolidated democracy; of course, Huntington insists, by the exemption of the Middle East and Muslim-based societies. The problem with terrorism denotes a much broader, deep-seated issue that represents the ongoing rejection of the Westernization process by Islam. These contrasting values are being disputed in terrorism issues (Huntington 1993, 1997).

At a closer look, Westernization and modernization are inevitably entwined. Although democracy shamefully surfaces in pre-Christian backgrounds, it is no less true that the US introduced a stable political system that paved the way for the invention of the check and balance institutions that prevent the advance of populism. With this in mind, Huntington says that,

> These countries tended to oscillate between more populist democratic governments and more conservative military regimes. Under democratic regime radicalism, corruption, and disorder reach unacceptable levels and the military overthrow it, to considerable popular relief and acclaim. In due course, however, the coalition supporting the military regime, unravels, the military regime fails to deal effectively with the country's economic problems, professionally inclined military officers become alarmed at the politicization of the armed forces, and again, to great popular relief and acclaim, the military withdraw from and are pushed out of office.
>
> (Huntington 1993: 42)

As Huntington contends, high levels of conflict exist even within democracies, but unlike totalitarian governments, democracy monopolizes the legal scaffolding to nuance violence and conflict. To put this bluntly, the most prosperous nations not only are democratic, but are less violent than authoritarian societies. This happens because the process of democratization played a leading role in the development of countries and economies (Huntington 1993). Among the numerous detractors of Huntington, we meet Edward Said. In his book *Orientalism*, he anticipates an answer to the question of Islamophobia, arguing that the East was considered as a threat which confronted the supremacy of the West (Said 1979, 1985). Said renames "clash of civilizations" as the "clash of ignorance", which denotes the lack of expertise of Huntington in Islamic issues. By this token, Said argues that the cultural representations around the Orient seem to be patronized by Europe under the label of "orientalism". This academic discipline not only echoes the long dormant colonial discourse, stereotypes and prejudices but also presents an image of the East as subordinated to European rationality. Orientalism, which is present in literary theory, exhibits an exaggeration of the difference disposed to nourish an ethnocentric discourse. This sentiment assumes a pre-fixed sentiment of superiority that results in an inaccurate diagnosis of Middle Eastern issues (Said 1979). The subordination of the East with respect to the West is centered on fictional landscapes, promoted by literature and travel writing, which persist to date. Terrorism, of course, aligns the existing Eurocentrism with previous beliefs about Arabs as radicalized, exuberant and irrational people. This happens because the European identity is reaffirmed according to the creation of an alter-ego, which legitimated Western domination. The knowledge available about the Eastern world coincides with a set of ideologies and fictions which were oriented to depict Arabs as "undemocratic", psychologically weak, or as children respecting the European status quo. Undoubtedly, Said was not only pivotal in post-colonial studies but also unveiled the compliance of scholarship with the ideological traps posed by the ruling elite. In the "Clash of ignorance", Said exerts a radical criticism of Huntington's thesis, detailing how he puts the cart before the horse. In fact, Said accepts that Huntington's argument rests on shaky foundations, using labels such as West, Democracy and Islam without further discussion. As stated, Said argues that one of the main limitations in Huntington's arguments is not only his lack of knowledge about Muslim culture, but also that he involuntarily embraces long-dormant stereotypes forged by colonial rule (Said 2001).

Said's accounts not only shed light on the cultural theory which unveiled the subtle discourses of Eurocentrism, but illuminated, for those scholars who focused their efforts on studying Islamophobia, the emergence of new, more subtle forms of "cultural" racism. Amir Saeed from the University of Sunderland alerts us to the fact that, not only are many Muslims segregated in the UK, but the image of Arabs is orchestrated as an ethnic group which remains outside the hosting lifestyle. Terrorism generally and 9/11 in particular reinforced a logic of fear and exclusion revolving around the Muslim faith. Arabs are portrayed not only as culturally inferior to Britons but as aliens who have been unilaterally inserted because of the countless humanitarian crises in Middle East. The Western media developed an uncanny trend to treat "Islam and the West as opposites and different. Although neither the West nor Islam exists as monolithic entities, journalists and politicians insist on framing the current situation in these terms" (Saeed 2007: 11). Other pictures most likely in films show Arabs as misogynist, violent or cruel. The Islamic community, according to the ideologists of Islamophobia, failed to be integrated with the European nations, living as outsiders who demand hospitality but without a genuine engagement with the hosting society. As Saeed eloquently observed, this image harms the reputation of long-established Arab communities at the same time as it destroys their integrity and cultural autonomy.

Islamophobia

For a great portion of the social imaginary, Muslims implicitly support the terrorist causes, not only facilitating them with lodgings or assistance in the suburban areas of the main European cities, but alluding to Islam as a religion based on anti-Western sentiment (Saeed 2005; Abbas 2012). Islamophobia, after all, exhibits the reactions of the Empire to regulate and control the flux of Muslim migration in Europe. 9/11 gives the perfect justification to treat Muslims and other non-Western aliens as subordinated to the cultural European matrix. As Saeed (2005) puts it,

> Since the events of 9/11 and the subsequent War on Terror, it could be argued that Muslim are now treated as the dominant threat of British society. Similar to other racialized groups, various authors contend that the media represent Muslims as one of homogenous uncivilized groups, with mass generalizations that often depict Muslims as terrorists or Islamic extremists.
>
> (Saeed 2005: 25)

Doubtless, around Islam the media has imposed a violent image, which is systematically replicated to protect the interests of the ruling classes. The stereotypes

of Muslims as terrorists have been replicated in almost the entire cultural entertainment industry, which ranges from films, to novels to video games. Islamophobia hides a great trauma in the post-9/11 era, reaching vast audiences across the
globe. The aftershocks generated by terrorism and the attacks on the World Trade
Center structured not only the internal politics of the US but also its international
relations policies (Awan 2010). Luke Howie studied the media impacts in distant
audiences, finding that the current power of amplification the media boasts resulted
in the expansion of a climate of fear and anxieties throughout the world (Howie
2015). In consonance with this, G. Skoll (2016) coins the term "global fear" to
denote how the American empire has systematically exploited two central figures:
the capital and the fear. While the former engenders some gaps between haves and
have-nots, the latter exerts an ideological control undermining social resistance
and dissidence. Equally important, consumers not only are cheated in view of the
human rights violations committed by their governments in the name of democracy
abroad, but they feel that capitalism should be esteemed as a wonderland. This ide-
alized world is placed in jeopardy by the advance of terrorism. Skoll concludes that
the imposition of fear takes different mechanisms and channels which have varied
throughout history. For example, Islamophobia today has continued the dynamics
of "the red-scare", the fear of domestic crime and the segregation of African-
Americans. The globalization of terror helps capital owners to immobilize unions
internally while terrorists are depicted as "enemies of democracy".

Graham Fuller (2010) brilliantly asks what would happen if you and me lived
in a world without Islam?

The West has stereotyped Muslims as a dangerous and blood-thirsty community, which cooperates with international *jihadist* terrorism. This led some scholars – like Huntington – to admit there are cultural incompatibilities between Islam
and Christianity. Fuller conceptually frames a more interesting thesis. If Islam had
never existed, most surely another civilization would occupy its place. In fact,
history has witnessed how the Abrahamic faiths, Christianity, Judaism and Islam,
coexisted in peaceful conditions. The East–West crisis has surely nothing to do
with religion. The unilateral interventions of the West in the Middle East paradoxically paved the way for the consolidation of Islam while, fundamentally,
it is divided into many factions, with different interpretations of Islam and cosmologies. As Fuller observed, the ignorance of American officials in their lack of
understanding about the complexity of the Muslim world not only resulted in the
tragedy of 9/11 but also in miscarried policies conducted by George Bush in Iraq
and Afghanistan. Huntington took the incorrect path confirming that civilizations
are programmed to collide. The fact is that the US received admiration for its technological capacities to expand, but this does not mean civilizations should clash.
From its inception, Islam was adapted to the local landscapes, negotiating with its
neighbors. Following Muhammad's death, Arabs entered into direct conflict with
other powers (like Rome) which had ambitions for military intervention in the
region. There is nothing like these cultural asymmetries between Islam and Christianity, to the extent that both are centered on the constituencies of the same values. Of course, if Islam never existed, the Byzantine Empire would occupy its role.

Islam, as a new geopolitical force, inherited not only much of the anti-Rome views that grew over time within Byzantine Empire itself. While Byzantium drew its deepest identity from the belief that it was perpetuating the true tradition of the Roman Empire, it increasingly came to view the Western Church as a geopolitical rival whose power was ultimately as threatening to Byzantine power and identity as Islam itself.

(Fuller 2010: 68)

Fuller's contributions help us to understand Islam as a historical political unit which was settled through the advance of Rome in the region.

One of the experts in the field, Bruno Etienne, fleshed out a very interesting model to situate Islam in context. Unlike the Catholic Church which coded a complex and strict dogma that teaches the lessons of Jesus to the community of believers, Muslims had no such luck. Not only did Muhammad's death surprise his followers, but from that moment on, there was a plethora of interpretations and schools revolving around the Koran (the most significant Muslim sacred text). As Etienne reminds us, not surprisingly the biblical texts sound ambiguous, misleading and contradictory. The Catholic Church centralized its authority on the basis of a unilineal interpretation of the Bible. The Koran, rather, is interpreted and lived from different angles and ways, which led Muslim communities in a richer but no less chaotic world. The concept of *jihad* has not the connotation given by Western analysts. The *jihad* represents the internal struggle of the soul to be a better person and change positively the environment. The *Umma* (community) comprises not only the political but the religious authority emanated by the Koran. When the *Umma* is at risk, Muslims declare the *jihad* against the intruders. Those who manifest their discontent with the established authority and do not agree with *jihad* are considered *Takfirs* (traitors). Though it sounds an over-simplification of the Koran, as Etienne admits, Islamic terrorism derives from biased lectures on the Koran and the lack of a centralized corpus indicating when and under what conditions the *jihad* may be invoked. While disunited or in disagreement with other kinships, the presence of an external power invites Muslims to form a unique force to repel the invasion. This occurred during the Soviet invasion and will repeat in the future (Etienne 1996, 2007; Cesari & Etienne 1997). Etienne's works illustrate the misinterpretation and distorted image of Islam on the part of the West. The specialized literature emphasizes how Muslim communities are symbolically represented as an outside (virus) encapsulated in the core of Western civilization (a healthy body). In the next section, we discuss critically to what extent 9/11 and the emerging fear of terrorism have shifted the ways of making politics in the US and Europe.

Discussing politics after 9/11

Among academics there is no consensus of precisely how 9/11 changed politics and day-to-day life. While some scholars catalogued the international *jihadism* as a serious threat to the West and to global peace (Gunaratna 2000; Keohane & Zeckhauser 2003; Sloan 2006), others hold the thesis that the discourse (culture) of fear instilled by terrorism is not news but a disciplinary mechanism in order for workers to accept economic (neoliberal) policies that otherwise would be widely neglected (Baudrillard 2003; Chomsky 2015; Bauman & Lyon 2013; Howie 2012; Skoll 2016; Korstanje 2017). On one hand, neo-pragmatists enthusiastically embrace the idea that the world, after the Soviet decline, transformed into a dangerous place, or at best a landscape which shows hostility against the US and democratic values (Revel 2003). The sentiment of anti-Americanism, which is fostered by some central European nations, grows with little resistance in under-developing countries or sites whipped by extreme poverty. Unlike neoliberals who accept free trade as a peacemaker instrument, neo-pragmatists strongly believe in the power of force (Curtis 2004). Hence, not only is terrorism implicitly seen as the tip of an iceberg, which is the Muslim invasion of Europe, but it justifies military intervention to protect the values of liberal democracy worldwide. In so doing, the idea of preventive attacks plays a leading role as a moral justification (Levy 1987; Goldstein 2006; Walzer 2015). Faced with an imminent act of aggression, the country reserves the right of self-defense anticipating a counter-attack on the intruders (Bellamy 2008). Fundamentally, this constitutes a paradoxical situation because the doctrine of preventive attacks contradicts the basic cultural values of liberalism, which historically struggled for the self-representation of nations. On the other hand, this polemic point inspired some left-wing writers such as Chomsky (2015) and Žižek (2008) to defy the would-be objectivity of neo-pragmatists. They are accused of being ideologists or "fear mongers" whose arguments are echoed and accordingly replicated by the voices of the most recalcitrant right-wing extremists on TV. For these scholars, terrorism does not represent any threat to the West but the psychological fear allows some ruling elites to overcome the check-and-balance powers that ensure the climate of equality and fraternity in a democracy. Since capitalism should be understood as a cultural and economic project aimed at creating inequalities in wealth distribution, it is no less true that the invention of an external(internal) enemy validates policies oriented to destroy the autonomy of workers and unions (Žižek 2008; Chomsky 2015; Skoll 2016). Then, to some extent terrorism and capitalism would be inevitably entwined.

Equally important, the last presidential debate between Hillary Clinton and Donald Trump synthetized two types of contrasting fears. At first glimpse, the "war on terror" involved Americans in a climate of fear, which paved the way for the rise of radicalized voices in politics (like Donald Trump). To some extent, the rage against aliens, the constructions of walls and the Muslim travel ban were some measures that harmed the autonomy of the democratic institutions.. However, it is important not to lose sight of the panic Americans feel in response to aliens. Donald Trump emulated the fear of strangers and Clinton the fear of tyranny. Doubtless, in the current times when terrorism threatens to destroy the European spirit of hospitality, not surprisingly the fear of strangers prevailed (Korstanje 2017).

In a powerful work, David Altheide (2017) presents an updated version of *Terrorism and the Politics of Fear*. He aims to explain how and why a radicalized voice like Trump reached the presidency. In a society of hyper-consumerism, critical thinking is a scarce commodity. Altheide adheres to the view that popular culture has commoditized terrorism not only as a form of entertainment, but as a political discourse that constrains the rights of some minorities (e.g., Muslims). In view of that, one might speculate that media packages, disseminates and replicates a political message behind its coverage. Media is not a passive actor, but one of the informational sources of modern politics. However, Altheide believes that far from what many scholars think, the actual climate of fear is not directly generated by terrorism.

Altheide acknowledges that fear is politically tergiversated for capitalism to eradicate those glitches that affect the process of production. In the name of security, capital-owners pose an economic decentralized background which destroys the rights of workers. Terrorism not only paralyzes critical thinking but paves the way for the advance of populism. The 9/11 attacks ignited a new era of "counterfeit politics", where conspiracy plots nourished surfacing paranoiac discourses. This facilitated a new politics of fear designed to elide the division of powers. The executive branch claimed to be the only one to lead the sacred fate of the nation, and for that, the other two branches should be silenced. In the midst of this mayhem, the Muslim community was and is daily demonized as an internal foe that at any time may cowardly attack the US. Beyond any responsible debate, 9/11 fed a false patriotism that ultimately seems to be economically unsustainable. Altheide contends that the new politics need victimization as a platform for an "evocative entertainment".

> Fear also makes us more compliant in seeking help or being rescued from formal agents of social control. This is very apparent with the rise of victimization as a status to be shared and enjoyed. Fear is a perspective or orientation to the world rather than fear of something. Fear is one of the few things that Americans share.
>
> (Altheide, 2017: 97)

In consonance with Altheide, this book calls attention to the incapacity of modern democracy to deter populism and racism. Above all, while the popular parlance manifests its fears on the advance of terrorism, the real effects of terrorism impacting on the democratic system are ignored. In fact, 9/11 fits perfectly in a society of the spectacle, cementing the start of an adversary democracy, where the theory of conspiracies and mistrust reign. In view of this, capitalism successfully disarticulated the social cohesion through a climate of competition and conflict (divide and rule tactic).

Complementarily, David Kelman (2012) sets out an alternative reading on politics. Based on literary theory, he argues convincingly that politics are often invigorated by the conspirational plot and a floating signifier, which is filled by the rulers. This does not mean that counterfeit politics derive from a pathological form of politics; the conspiracy rather was rooted in the ideological core of politics. For some reasons, social scientists of all stripes explored the limitations of ideology and its intersection in politics, but less attention was given to the role of conspiracy in society. The silence, as Kelman insists, produces a gap between the formal and the informal story about traumatic events. This crosses the cultural differences, from Latin to Anglo-America. Historically, modern democracies appealed to the stimulation of conspiracy in order for the ruling class to gain further credibility. Unlike Altheide, Kelman writes that,

> politics is not based on an ideology decided in advance, but it is rather constituted through a specific type of narrative that is often called conspiracy theory. This type of theory is always a machination, that is, a narrative mechanism that secretes, as it were, ideological labels such as the right or the left.
> (Kelman 2012: 8)

The floating signifier creates two contrasting bands, them and us, dialoguing in a confrontational logic. Each conspirational narrative contains a double-standard structure in which the official story constantly confronts fictionality. Kelman acknowledges that the conspiracy seems not to be the symptom of a political crisis but the nature of politics. To put this slightly in other terms, not only does Kelman's project contrast the argument of classical political theory by posing politics in connection with secrets, but it opens the doors for a paradoxical situation. When a social and political system is reproduced, communities face a threatening event. This simply happens because politics occurs when one discourse is being undermined by another contrasting voice. "In other words, conspiracy theories willfully suspend the normal procedures of politics in order to combat an internal enemy that is threatening to destroy the stability of the polis" (Kelman 2012: 123).

Comparing conspiracy to detective novels, Kelman defines "counterfeit politics" as the urgency of rediscovering (domesticating) the darkness, the secret. Paradoxically, like the plots in novels, the detective traces and investigates to discover a crime, and of course, his or her endeavors are placed in elucidating who was the killer, but in so doing, the detective is trapped into a gridlock because the crime finally never happens. This means that politics are fostered and backed by the same confrontational instability it generates (Kelman 2012).

Though thoughtful, inspirational and self-explanatory, Kelman's argument rests on a serious caveat. He overlooks that conspiracy and ideology are two sides of the same coin. When the ideology (as a system of ideas and beliefs aimed at explaining reality) is not enough, the theory of conspiracy gives a distorted version of the facts. The US is portrayed by the film industry as the most powerful nation in the world, the watchdog of an ever-conflicting world. That the attacks on the World Trade Center happened in 2001 not only contradicted this, but evinced that, after all, the most powerful nation may be under attack. The power of ideology exaggerated the image of the US while the capitalist productive system was exported as a triumphant model of prosperity (Korstanje 2018).

The last of the travels

In the Western social imaginary there is a strange paternalist discourse aimed at presenting capitalism as the best of all feasible worlds. In fact, the non-Western "Other" is ambiguously accepted but frightens us as a potential danger. This moot point will be debated in this section through the novel authored by Eric Frattini.

Frattini has vast experience as a journalist as well as a war correspondent in the Middle East. His novels and texts have been translated into more than a dozen languages and, over recent years, have focused on the efforts of Western governments to defeat terrorism. This is his main concern, found throughout his novel *La Lenta Agonia de los Peces* (*The Slow Agony of Fishes*), which was published by ESPASA in 2013. The plot centers on the life of fictional character Havana Sinclair, who is a brilliant counter-terrorist agent in the service of British intelligence. Although she is one of the most prominent experts in Islamic terrorism who always sees matters through, she is fired from her unit, MI5, after the real-life 7/7 London bombings that occurred in 2005. Havana meets a private insurance company that requested her experience and services in Baghdad. She is asked to investigate the murders of two millionaire investors. Once there, she struggles against old foes.

The story situates around a real-life terrorist attack in London, England, where four bombings killed 52 civilians. The first explosion occurred at 8.50 between Liverpool Street Station and Aldgate, while two more were detonated at Edgware Road and Russell Square. The last attack was on a double-decker bus in Tavistock Square. One of the terrorists was Hasib Hussain, a British Muslim whose parents came from Pakistan. His father served as a rank-and-file factory worker and his mother aided newcomers and migrants with their English. Not only was Hasib lucky enough to go to high school, but he was also known as a smart boy, very polite and gifted. Here the question is, what happened to make Hasib become a terrorist?

Whatever the answer may be, the investigation indicated that he was voluntarily involved at a mosque where he met with some radical terrorists. In parallel, in the novel, Havana is fired from her MI5 position. What is more interesting in Frattini's argument is that not only was London selected as the host city for the 2012 Olympic Games the day before the bombings, but was also selected by terrorism. These four strikes left 52 victims and more than 700 wounded. Such a traumatic event also cost Havana Sinclair her position as an agent and terrorism expert. As an incorruptible woman, Havana travels to Baghdad to investigate these crimes. She moves in the shadows of British intelligence, interrogating herself about the extent to which she has chosen the right side. Frattini, who has served as a well-known expert in terrorism, imagined a fiction, which has been favorably compared to reality.

The novel gradually ushers readers into the world of Islamic terrorism, combining some aspects of fictionality with the current challenges Europe faces today. The fear of terrorism is accompanied by a deeper mistrust of government. Frattini's novel is centered on an ethnocentric discourse that depicts the Muslim community, or, to be more precise, British Muslims, as potential terrorists. The novel ideologically portrays Hasib as a man who has everything his parents lacked. He is smart, born and well-educated in a developed country, but this does not suffice for him. In the quest for a more authentic lifestyle, he needs to get closer to religion to develop an intricate hatred of the UK. This raises an extremely interesting question: is religion the key factor in Muslim terrorism?

In contrast with Hasib, Havana is a mere British citizen, who not only sacrifices her life for her nation, but also fights against dark powers to protect homeland security without asking anything in return. She is blamed for what happened but she still does what should be done. Once in Iraq, where the American army has deployed thousands of soldiers, she meets Jalid, who will be her driver and escort. This city is dangerous for everyone, and particularly for women. Havana is finally kidnapped by a commando group in the Mutanabbi district. Frattini (2013) portrays Iraq as a grim landscape, a country devastated by war.

> Havana, todavía con la capucha puesta sobre la cabeza, olía los medados y excrementos que se amontonaban a su lado. Las moscas se posaban sobre su piel y su ropa, pagada al cuerpo por el sudor, ante la atenta mirada de un adolescente que la vigilaba con un rifle sobre sus piernas. [Havana, hooded on her head smelt the piss and shit accumulated along her, while flies landed on her skin and clothing, which was stuck to her body by the extreme sweat. Before her, a teenager kept a gun between his legs.]
>
> (Frattini 2013: 153)

Like many other teenagers in Iraq – for Frattini – this teenager shares a similar story, living in a devastated country; doubtless, he is a victim but at the same time, a killer who assassinates on the orders of maniacal leaders. Iraq is a site of war and devastation, where children have no future. His father was executed by the former regime and his two brothers were unjustly killed by the current government. As an insurgent, the boy knows one of his brothers was jailed at Abu Ghraib while the other lost his life as a suicide bomber at the Wall of Zona Verde (Green Zone). Throughout Havana's captivity, she realizes how the ideals of human rights, as they are imagined in the West, are systematically violated in the periphery. In this respect, Havana is held with a contractor, who tells her that they will all die. She answers that they have no reason to think this, but the trembling businessman tells her:

> Soy contratista, sabes que significa esto en Irak? Desde que nos capturaron a Charlie y a mi, me han humillado y me han torturado". [I am a contractor, do you know what that means in Iraq? Since Charlie and I have been kidnapped by these guys we have been humiliated and tortured.]
>
> (Frattini 2013: 174)

Though Havana is confident that she will be liberated, during their long conversations the contractor tells her that he is a hitman contracted to target some specific persons. This secret surprises Havana, who suspects that after all, it is no coincidence that he is there. He was commanded to assassinate two hunters, Tibbals and McMillan, who loved hunting human beings as trophies for sport. Havana is unsure whether she will live, but what is most important, she sees the grim future of Iraqis, creating a sentiment that leads her to confront her own Western presumptions. She is finally released by an unknown person and found in the street by a British patrol. She misses her husband (Peter Sinclair), who represents the British oligarchy, and her daughter. She is faced with serious dangers and risks on her return to England. After all, those who have carefully planned the attacks in London were monitoring her steps. Havana not only discovers a dark conspiracy planned by high-ranking British intelligence officials such as Raymond Clarke, but she realizes the plans were orchestrated to change the internal politics of the United Kingdom. In his novel, Frattini depicts a typical landscape. Some corrupt agents located at the top of the pyramid of the security forces or the government who should care for citizens instead collude with Muslim terrorists to encourage a new social order.

To cut a long story short, *La Lenta Agonia de los Peces* presents as a well-written novel with the main theme of terrorism. At first glimpse, Frattini follows an ethnocentric discourse that has the following elements: a strong and incorruptible hero (Havana) who sacrifices her life and comfort, constituent of the developed world to protect others. This sacrifice takes her to remote and lawless nations where human rights are violated daily and where people die for no reason. The pre-Hobbesian state of chaos and violence that characterizes Iraq, which is a zone occupied by the Anglo-alliance, seems to be one of the grim landscapes Frattini often draws. On the other hand, Havana's husband emulates the passivity of the elite, which declares itself unable to fight against terrorism. The perpetrators of the London bombings were first-generation British Asians whose parents were welcomed under the sacred law of hospitality. Hasib appeared to be a brilliant and promising student who embraced a sentiment of hate and radicalization against the UK, risking in this way his opportunity to have a better life. Here we find some of Howie's comments respecting the unsaid terror, which is shaped by the articulation of countless complex narratives. Terrorists live like us, they have infiltrated the core of Western nations, they may even be our neighbors or, worse, friends. This puts us all in a difficult position because anyone may be a victim of terrorism, anytime and anywhere. As discussed in earlier chapters, terrorists target spaces of mobility and consumption not only to humiliate the West, but also to show how the modes of mobility can be destroyed.

Conclusion

The present chapter investigated not only the growing decomposition of the social fabric, which paved the way for the rise of conspirational plots, but also a new culture of entertainment that commoditizes terrorism as a form of spectacle. Through the articulation of different ideological narratives, the ruling elite imposed a biased image of the Muslim community as a Trojan horse. As outsiders, Muslims are daily framed as dangerous to the Western lifestyle. The liberal state, from its outset, is programmed to discipline and inoculate against external dangers (paraphrasing Foucault) in the same way a vaccine is obtained after dispossessing the virus of all its destructive features. However, the turn of the twenty-first century liberated some dormant fears against the non-Western "Others". Not fully integrated with the local community like in Latin America, since the Syrian migration dating back to the nineteenth century, in the Global North, Islam is an object of fear and anxiety. Echoing the metaphor of the medical gaze, which is designed to locate and eradicate disease, Western rationality does not understand the gray zones. We used the example of cancer, which is potentially seen as a degeneration of cells. Western rationality will look at saving the body, extirpating the affected organ if necessary. This appears to be the main reason why the Muslim community runs a serious risk unless contingency plans are orchestrated against terrorism.

This book was oriented to discuss not only the socio-economic background of terrorism but also its effects, unearthing old lessons that Latin Americans remember. Terrorism is centered on a logic of instrumentalization, which is masked in the heart of Western rationale. Democracy, by this token, takes different shapes according to the invocations of the political parties. Through this book we adamantly described and discussed the different theories around democracy and explained how terrorism may place it in jeopardy. We reviewed the origin and evolution of liberal doctrine, which was functional to the creation of the liberal state to understand the world of consumption. We pitted Zygmunt Bauman's contributions against the arguments of some liberal scholars such as Kathleen Donohue. Besides, we placed the theory of gaze under a critical lens reminding us that terrorism has been transformed through a conduit in a society that valorizes the Others' suffering over other cultural values. Finally, in Chapters 4, 5 and 6, we revised and offered an updated discussion revolving around the role of the State in enforcing the law and national security, and the ethical limitations of torture. In fact, as the sad story of the five Argentinian tourists killed in New York shows, terrorism involves two important assumptions.

Firstly and most importantly, the colonial project developed a restricted hospitality giving the alterity an inferior status to that of the European matrix. Through the travels and the importation of leisure activities, nation-states gradually constructed a complex net of exchange and trade that positioned Europe at the top of the developed nations. However, the imperial enterprise acquired a strange aversion to aliens. The myth of the Trojan horse can be applied to the treatment of Muslims in the developed world. They are objects of mistrust and fear, fertilizing the idea that terrorism looks to destroy democracies. We discussed critically the roots of ancient democracy and its mutation to Anglo-democracy, as well as the role of torture to get vital information in an emergency context.

Secondly, we paid attention to the signs of decomposition which is evinced by the rise of counterfeit politics. Most likely the best example that reveals how democracies are ideologically invoked to protect their proper interests is the case of Catalunya. While Carles Puigdemont, a Catalan Nationalist and President of Catalunya, declared the independence of the central administration hosted in Madrid, Mariano Rajoy catalogued the declaration as an act of treason and sedition. It created an escalation of verbal accusations from one to the other side, which resulted in the exile of Puigdemont and fierce interventions in the popular referendum for independence. Catalans, echoing Puigdemont, wanted to vote, accessing their right to democracy, while those loyalists to Spain called on democracy and the law as the touchstone of the Republic. Both sides, after all, invoked a different face of democracy (as a floating signifier) manipulated according to local convenience. What is most important, Catalunya claimed independence without fiscal autonomy (capital) or armed forces (the repressive apparatus of State). The rest of the story is well known; Puigdemont and the separatists' dreams withered away in view of a process of feudalization, accelerated after the last stock market crisis, which produced radical ruptures between factions and provinces.

Quite aside from the answers for the Catalan case, this book tackled the problem of terrorism and its impact in the fields of democracy, mobility, multiculturalism and the consumerist lifestyle. The different chapters forming this book were aimed at showing how terrorism decomposes the social ties, which are vital for the social fabric, affecting one of the symbolic touchstones of Western civilization: *hospitality*.

References

Abbas, T. (2012). The symbiotic relationship between Islamophobia and radicalisation. *Critical Studies on Terrorism*, *5*(3), 345–358.

Allport, G. W. (1979). *The Nature of Prejudice*. New York, Addison Wesley.

Altheide, D. (2017). *Terrorism and the Politics of Fear*. New York, Rowman & Littlefield.

Awan, S. M. (2010). Global terror and the rise of xenophobia/Islamophobia: An analysis of American cultural production since September 11. *Islamic Studies*, *29*(4), 521–537.

Baudrillard, J. (2003). *The Spirit of Terrorism and Other Essays*. London, Verso.

Bauman, Z., & Lyon, D. (2013). *Liquid Surveillance: A Conversation*. New York, John Wiley & Sons.

Bellamy, A. J. (2008). The responsibilities of victory: Jus post bellum and the just war. *Review of International Studies*, *34*(4), 601–625.

Borradori, G. (2013). *Philosophy in a Time of Terror: Dialogues with Jurgen Habermas and Jacques Derrida*. University of Chicago Press.

Cesari, J., & Étienne, B. (1997). *Etre musulman en France aujourd'hui*. [*Being Muslim in France Today*.] Paris, Hachette.

Chomsky, N. (2015). *Culture of Terrorism*. New York, Haymarket Books.

Curtis, M. (2004). Anti-Americanism in Europe. *American Foreign Policy Interests*, *26*(5), 367–384.

Eid, M. (Ed.). (2014). *Exchanging Terrorism Oxygen for Media Airwaves: The Age of Terroredia*. Hershey, IGI Global.

Elias, N., & Scotson, J. L. (1994). *The Established and the Outsiders* (Vol. 32). London, Sage.

Etienne, B. (1996). *El islamismo radical (The Radical Islamism)*. Madrid, Siglo XXI de España Editores.

Etienne, B. (2007). Islam and violence. *History and Anthropology*, *18*(3), 237–248.

Foucault, M. (2003). *Society Must Be Defended: Lectures at the Collège de France, 1975–1976* (Vol. 1). London, Macmillan.

Frattini, E. (2013). *La Lenta Agonia de los Peses*. [*The Slow Agony of the Fishes*.] Barcelona, Espassa Narrativa.

Fuller, G. E. (2010). *A World Without Islam*. New York, Little Brown.

Goldstein, L. (2006). *Preventive Attack and Weapons of Mass Destruction: A Comparative Historical Analysis*. Stanford, Stanford University Press.

Gunaratna, R. (2000). Suicide terrorism: A global threat. *Janes Intelligence Review*, *12*(4), 52–55.

Howie, L. (2012). *Witnesses to Terror: Understanding the Meanings and Consequences of Terrorism*. New York, Springer.

Howie, L. (2015). Witnessing terrorism. *Journal of Sociology*, *51*(3), 507–521.

Howie, L., & Campbell, P. (2017). *Crisis and Terror in the Age of Anxiety: 9/11, the Global Financial Crisis and ISIS*. New York, Springer.

Huntington, S. P. (1993). *The Third Wave: Democratization in the Late Twentieth Century*. Oklahoma, Oklahoma University Press.

Huntington, S. P. (1997). *The Clash of Civilizations: Remaking of World Order*. New York, Touchstone Books.

Jerónimo, M. B., & Monteiro, J. P. (Eds). (2017). Past to be unveiled: the interconnections between the international and the Imperial. *Internationalism, Imperialism and the Formation of the Contemporary World: The Pasts of the Present*. New York, Springer, 1–32.

Kelman, D. (2012). *Counterfeit Politics: Secret Plots and Conspiracy Narratives in the Americas*. Maryland, Bucknell University Press.

Keohane, N. O., & Zeckhauser, R. J. (2003). The ecology of terror defense. *The Journal of Risk and Uncertainness*, *26*(2–3), 201–229.

Korstanje, M. (2017). *Terrorism, Tourism and the End of Hospitality in the West*. New York, Springer Nature.

Korstanje, M. E. (2018). *Tracing Spikes in Fear and Narcissism in Western Democracies Since 9/11*. Oxford, Peter Lang.

Levy, J. S. (1987). Declining power and the preventive motivation for war. *World Politics*, *40*(1), 82–107.

O'Malley, A. (2017). "An Anvil of Internationalism: The United Nations and Anglo-American Relations during the Debate over Katanga 1960–1963". In *Internationalism, Imperialism and the Formation of the Contemporary World: The Pasts of the Present*, M. B. Jerónimo & J. P. Monteiro (Eds). New York, Springer, 127–146.

Poole, E. (2002). *Reporting Islam: Media Representations and British Muslims*. London, IB Tauris.

Rajan, R. G. (2011). *Fault Lines: How Hidden Fractures Still Threaten the World Economy*. Princeton, Princeton University Press.

Revel, J. F. (2003). *Anti-Americanism*. San Francisco, Encounter Books.

Saeed, A. (2005). Racism and Islamophobia: A personal perspective. *Identity Papers: A Journal of British and Irish Studies, 1*(1), 15–31.

Saeed, A. (2007). The representation of Islam and Muslims in the media. *Sociology Compass, 1*(1), 1–13.

Said, E. (1979). *Orientalism*. New York, Vintage.

Said, E. W. (1985). Orientalism reconsidered. *Race & Class, 27*(2), 1–15.

Said, E. W. (2001). The clash of ignorance. *The Nation*, 22.

Skoll, G. R. (2016). *Globalization of American Fear Culture: The Empire in the Twenty-First Century*. New York, Springer.

Sloan, S. (2006). *Terrorism: The Present Threat in Context*. Oxford, Berg.

Tajfel, H. (1969). Cognitive aspects of prejudice. *Journal of Biosocial Science, 1*(S1), 173–191.

Tajfel, H. (1981). *Human Groups and Social Categories: Studies in Social Psychology.* Cambridge, Cambridge University Press.

Van Dijk, T. A. (1992). Discourse and the denial of racism. *Discourse & Society, 3*(1), 87–118.

Van Dijk, T. A. (1993). *Elite Discourse and Racism* (Vol. 6). London, Sage.

Van Dijk, T. A. (2000). New(s) racism: A discourse analytical approach. *Ethnic Minorities and the Media, 37,* 33–49.

Walzer, M. (2015). *Just and Unjust Wars: A Moral Argument with Historical Illustrations.* New York, Basic Books.

Žižek, S. (2008). *Violence: Six Sideways Reflections.* London, Verso.

Index

Printed and bound by CPI Group (UK) Ltd, Croydon, CR0 4YY

24/10/2024

01778282-0017